Recent Progress, Challenges and Outlooks of Insulation System in HVDC: 2nd Edition

Recent Progress, Challenges and Outlooks of Insulation System in HVDC: 2nd Edition

Guest Editors

Chuyan Zhang
Xiaobo Meng
Hao Yang
Zhong Wang

MDPI

Basel • Beijing • Wuhan • Barcelona • Belgrade • Novi Sad • Cluj • Manchester

Guest Editors

Chuyan Zhang
School of Artificial
Intelligence
China University of
Geosciences Beijing
Beijing
China

Xiaobo Meng
School of Mechanical and
Electrical Engineering
Guangzhou University
Guangzhou
China

Hao Yang
School of Electronics and
Information
Xi'an Polytechnic University
Xi'an
China

Zhong Wang
College of Electrical
Engineering
Sichuan University
Chengdu
China

Editorial Office
MDPI AG
Grosspeteranlage 5
4052 Basel, Switzerland

This is a reprint of the Special Issue, published open access by the journal *Energies* (ISSN 1996-1073), freely accessible at: https://www.mdpi.com/journal/energies/special_issues/QV188VQ5G7.

For citation purposes, cite each article independently as indicated on the article page online and as indicated below:

Lastname, A.A.; Lastname, B.B. Article Title. *Journal Name* **Year**, *Volume Number*, Page Range.

ISBN 978-3-7258-6972-5 (Hbk)
ISBN 978-3-7258-6973-2 (PDF)
https://doi.org/10.3390/books978-3-7258-6973-2

Contents

energies

Editorial

Recent Progress, Challenges and Outlooks of Insulation System in HVDC: A Further Discussion

Chuyan Zhang [1,*], Hao Yang [2], Xiaobo Meng [3] and Zhong Wang [4]

1 School of Artificial Intelligence, China University of Geosciences Beijing, Beijing 100083, China
2 School of Electronics and Information, Xi'an Polytechnic University, Xi'an 710048, China; yanghao@xpu.edu.cn
3 School of Mechanical and Electrical Engineering, Guangzhou University, Guangzhou 510006, China; mengxb@gzhu.edu.cn
4 College of Electrical Engineering, Sichuan University, Chengdu 610065, China; zhongwang@scu.edu.cn
* Correspondence: zcy@cugb.edu.cn

By 2025, China successfully commissioned more than 40 Ultra-High Voltage (UHV) power transmission and transformation projects, establishing the world's largest and most technologically advanced UHV grid. The operational footprint of this network in terms of energy resource dispatching and allocation spans the entire nation. Driven by the mandate for socio-economic sustainable development, the proliferation of renewable energy integration has necessitated high-voltage infrastructure, which effectively mitigates the curtailment of intermittent power generation and resolves the geospatial mismatch between generation sources and load centers.

Especially in Southern China, numerous UHVDC (Ultra-High Voltage Direct Current) transmission projects serve as critical corridors delivering bulk clean hydropower to megacity load clusters. These large-capacity projects traverse distances exceeding 1000 km, operating in harsh geographical environments characterized by complex atmospheric conditions. Ensuring the secure and stable operation of these capital-intensive assets is inextricably linked to maintaining the integrity and dielectric performance of the insulation systems within an optimal state.

Having this in mind, we are pleased to launch the second edition of the Special Issue titled "Recent Progress, Challenges and Outlooks of Insulation System in HVDC." This initiative is dedicated to disseminating state-of-the-art research findings regarding insulation coordination, failure mechanisms, and condition monitoring in HVDC transmission engineering, with the aim of fostering extensive and in-depth academic exchange among researchers.

The second edition of this Special Issue comprises a total of 10 contributions, including eight research articles and two review papers. The scope of these studies encompasses primary insulation equipment critical to HVDC engineering, such as circuit breakers, transmission line insulators, conductors, high-voltage bushings, power cables, and Gas Insulated Switchgear (GIS). The contributing authors present cutting-edge research findings addressing diverse insulation structures and characteristics, as well as the application of novel methodologies for the assessment and analysis of insulation performance.

Specifically, in Contribution 1, Hongping Shao et al. propose a novel detection technique for combined faults in circuit breaker operating mechanisms based on Deep Residual Networks (ResNet). Addressing the complex mechanical structure and potential failure sources of circuit breakers, this study innovatively incorporates spring-damping elements into the simulation analysis, thereby enabling the model to reflect real-world fault scenarios with higher fidelity. Furthermore, to tackle the challenge of small-sample data associated with mechanism faults, the research utilizes the ResNet50 architecture for spectrum data processing. By integrating the ReLU (Rectified Linear Unit) activation function, the proposed method achieves a fault identification accuracy exceeding 90%.

Moreover, the authors in paper Contribution 2 propose a methodology for diagnosing zero-value insulators on transmission lines. As zero-value insulators pose a long-term threat to the reliability of line insulation, this study leverages high-fidelity simulation results to develop a model based on Multi-Layer Perceptron (MLP) neural networks, capable of predicting the electric field intensity of defective zero-value insulators. Furthermore, a database characterizing the spatial electric field distribution of insulator strings containing zero-value defects was established, which has significant practical implications for the inspection and maintenance of external line insulation.

When it comes to the corona phenomenon on HVDC line conductors, Jules Simplice Djeumen et al. investigate the characteristics of space charge generated by DC corona through both experimental and simulation approaches Contribution 3. This study effectively integrates experimental data obtained from a corona cage with simulation results derived from Finite Element Method (FEM) models, enabling a comprehensive assessment of DC corona characteristics under varying ambient temperatures. Utilizing an indoor corona cage, the study tested different temperature gradients on standard models representing potential HVDC transmission lines in Southern Africa, acquiring critical data such as the Corona Inception Voltage (CIV).

Article Contribution 4 investigates the insulation reliability of epoxy resin-impregnated bushings for power transformers. Solid-insulation transformers have garnered significant attention in recent years. As critical insulation components within power transformers, epoxy resin-impregnated paper bushings are subjected to complex coupled multi-physical fields over the long term. Given the high manufacturing costs of these components, simulation analysis serves as an effective means to verify their insulation reliability. In this study, Daijun Liu et al. employ simulation to investigate the electric field distribution, thermal field distribution, and seismic performance verification of epoxy resin-impregnated paper bushings. In the study, the authors determine the maximum radial and axial electric field intensities for bushings of three different voltage levels, conduct thermal stability analysis under short-circuit conditions, and confirm that the bushing prototypes can effectively withstand seismic loads up to seismic intensity VIII, thereby satisfying product design requirements.

As the scale of the power grid expands, its structural complexity increases, and AC–DC hybrid grids inevitably face challenges regarding the interaction between HVDC and HVAC transmission lines. To address this, Jinyuan Xing et al. analyze the ground electric field of hybrid transmission lines Contribution 5. By adopting the preconditioned Krylov subspace method to solve discretized equations, the authors effectively calculate the spatial electric field when DC and AC components are coupled. This research provides valuable practical recommendations for determining critical design parameters—such as the optimal transmission tower configuration, minimum heights for towers and overhead conductors, and transmission corridor width—under various electromagnetic constraints.

Power cables constitute a vital mode of power transmission; the advancement of HVDC cables is constrained by the properties of insulation materials, with dielectric interfaces often representing the weakest link in the insulation system. Accordingly, Article Contribution 6 focuses on the performance of insulation interfaces in HVDC cable accessories. The authors propose a novel interface coating agent based on Polycyclic Aromatic Hydrocarbon (PAH)-modified silicone oil to enhance the insulation performance of HVDC cable accessories. This study investigates the DC breakdown characteristics of the insulation

https://doi.org/10.3390/en19030814

dielectric interface under conditions of corona and thermal cycling aging. By integrating results from breakdown tests, infrared spectroscopy, and microstructural analysis, the insulation aging mechanisms are elucidated. The research reveals that silicone oil modified with PAHs can significantly increase the breakdown voltage. Furthermore, Quantum Chemical Calculations (QCCs) identified 2,4-dihydroxybenzophenone ($C_{13}H_{10}O_3$) as the optimal additive for the interface coating.

Furthermore, Xiaolai Li et al. investigated the impact of a unique form of insulator surface contamination—"green algae"—on the surface characteristics of porcelain insulators Contribution 7. In the warm and humid mountainous regions of Southwest China, algae adherence to DC insulators significantly affects surface wettability, thereby increasing the risk of flashover under extreme weather conditions. This study proposes a method to measure the water absorption of the contamination layer based on surface conductivity, which enables the quantitative evaluation of the algae's influence on surface wettability characteristics. The results indicate that naturally deposited pollution has a negligible effect on the saturated water absorption of the contamination layer, whereas the presence of algae significantly alters the wetting behavior, leading to an increase in saturated water absorption. This finding provides profound insights into the role of biological contamination in the pollution flashover process of insulators in high-humidity regions.

Paper Contribution 8 investigates the mechanisms of electric field enhancement and interface synergistic effects in the insulation failure of SF_6/epoxy resin insulation systems. Through experiments conducted in SF_6 gas gaps with a length of 36 mm and pressures ranging from 0.1 to 0.4 MPa, the study measured the AC breakdown voltages of different post insulator samples under Dielectric-Initiated Breakdown (DIBD) and Electrode-Initiated Breakdown (EIBD), thereby analyzing the critical role of the solid dielectric surface during the initial stages of gas–solid interface discharge. Experimental results indicate that, under comparable maximum electric field (E_{max}) conditions, the AC breakdown voltage required for DIBD is significantly lower than that for EIBD. This phenomenon can be attributed to three primary factors: the modulation of electric field distribution by the dielectric material and shielding electrodes; the fact that microscopic irregularities on the dielectric surface induce stronger local electric field enhancement compared to similar features on metal electrodes; and the desorption process near high-field regions adjacent to the dielectric surface, which significantly enhances electron multiplication during the gas discharge process. This study enriches the understanding of interface insulation performance in gas-insulated systems and provides valuable references for engineering insulation design.

The final papers, Contributions 9 and 10, are review papers. In paper Contribution 9, Haohua Hu et al. systematically review and summarize the breakdown characteristics and discharge mechanisms of overhead transmission line gaps under vegetation fire conditions. HVDC transmission lines traversing forest areas are commonplace, and wildfires pose a significant threat to the operational stability of these lines. This study reviews extensive research on gap breakdown tests and discharge mechanisms in simulated wildfire conditions, while also analyzing and summarizing physical parameter measurement methods commonly used in current experiments. Furthermore, the paper clarifies the breakdown characteristics and discharge mechanisms derived from existing experiments and numerical simulations under various influencing factors, highlighting the applicability and limitations of these findings. In the end, the authors provide an outlook on future work in this research direction.

As for paper Contribution 10, the authors review the application of digital twin technology in the full lifecycle management of high-voltage bushings. The study defines a five-dimensional digital twin framework encompassing the entire lifecycle of high-voltage power equipment (design, manufacturing, operation and maintenance, and decommis-

https://doi.org/10.3390/en19030814

sioning). It deeply explores the application paradigms of digital twins in typical scenarios and categorizes core enabling technologies, including multi-physics coupled modeling, multi-source heterogeneous data fusion, and data-driven model updating and condition assessment. In the last part, the paper identifies the current challenges facing digital twins for high-voltage power equipment in terms of data, models, standards, and costs, and provides an outlook on future research directions.

Finally, I would like to express my sincere gratitude to all the authors for their contributions and cooperation, as well as extend my special appreciation to the editorial staff and reviewers for their dedicated efforts. Insulation performance is a critical attribute of high-voltage power equipment. Currently, with the progressive integration of technologies such as big data, artificial intelligence (AI), the Internet of Things (IoT), and advanced sensing into high-voltage power equipment, the design, installation, operation, maintenance, and management of power infrastructure are continuously evolving towards digitalization and intelligence. Consequently, we will place greater emphasis on research achievements in this field in the future, aiming to provide enhanced opportunities and platforms to foster more extensive and effective academic exchange.

Conflicts of Interest: The authors declare no conflicts of interest.

List of Contributions:

1. Shao, H.; Jiang, Y.; Zhao, J.; Li, X.; Zhang, M.; Yang, M.; Wang, X.; Yang, H. Research on Fault Detection Technology for Circuit Breaker Operating Mechanism Combinations Based on Deep Residual Networks. *Energies* **2025**, *18*, 1154.
2. Zheng, L.; Yin, P.; Li, J.; Liu, H.; Li, T.; Luo, H. Research on Diagnostic Methods for Zero-Value Insulators in 110 kV Transmission Lines Based on Spatial Distribution Characteristics of Electric Fields. *Energies* **2025**, *18*, 1534.
3. Djeumen, J.S.; Langa, H.M.; Sutherland, T. Corona-Generated Space Charge Characteristic in an Indoor HVDC Corona Cage Under Atmospheric Temperature Conditions. *Energies* **2025**, *18*, 2872.
4. Liu, D.; Tong, X.; Liu, L.; Dong, X.; Yan, T.; Tang, W.; Wang, L.; Cao, B.; Luo, Z. A Simulation and a Computational Study on the Reliability Verification of Epoxy Resin Paper-Impregnated Bushings in Power Transformers. *Energies* **2025**, *18*, 3239.
5. Xing, J.; Han, C.; Tian, J.; Wu, H.; Lu, T. Influence of Ion Flow Field on the Design of Hybrid HVAC and HVDC Transmission Lines with Different Configurations. *Energies* **2025**, *18*, 3657.
6. Zhu, W.; Pathiraja, K.; Guo, X.; Hui, B.; Fu, M.; Zhao, L.; Wang, Y.; Li, J. Electrical–Thermal Aging Performance of PAH-Modified Interfacial Coating Agent for HVDC Cable Accessory. *Energies* **2025**, *18*, 3767.
7. Li, X.; Wu, X.; Yang, S.; Gao, B.; Li, L.; Cao, B. Influence of Green Algae on the Surface Wetting Characteristics of Porcelain Insulators. *Energies* **2025**, *18*, 4212.
8. Liu, L.; Zhang, Q.; Peng, X.; Li, X.; Wang, Z.; Yu, S. Pressure-Dependent Breakdown Voltage in SF6/Epoxy Resin Insulation Systems: Electric Field Enhancement Mechanisms and Interfacial Synergy. *Energies* **2025**, *18*, 5014.
9. Hu, H.; Li, P.; Huang, D. The Progress and Prospect of Gap Breakdown Characteristics and Discharge Mechanisms of Overhead Transmission Lines Under Vegetation Fire Conditions. *Energies* **2025**, *18*, 1946.
10. Chi, W.; Wang, T.; Zhang, J.; Wang, Z.; Zhang, C. Full-Life-Cycle Management of High-Voltage Bushings Based on Digital Twin: Typical Scenarios, Core Technologies, and Research Prospects. *Energies* **2025**, *18*, 6343.

https://doi.org/10.3390/en19030814

energies

Article

Research on Fault Detection Technology for Circuit Breaker Operating Mechanism Combinations Based on Deep Residual Networks

Hongping Shao [1], Yizhe Jiang [1], Jianeng Zhao [1], Xueteng Li [1], Mingzhan Zhang [1], Mingkun Yang [2], Xinyu Wang [3] and Hao Yang [3,*]

[1] Dali Power Supply Bureau of Yunnan Power Grid, Dali 671000, China; 18691348636@163.com (H.S.)
[2] Electric Power Research Institute of Yunnan Power Grid Co., Ltd., Kunming 650220, China
[3] School of Electronic and Information, Xi'an Polytechinc University, Xi'an 710048, China
* Correspondence: yanghao@xpu.edu.cn

Abstract: Due to the complex mechanical structure of the spring-operated mechanism, its failure mechanisms often exhibit a multi-faceted nature, involving various potential failure sources. Therefore, conducting a failure mechanism analysis for multi-source faults in such systems is essential. This study focuses on the design of composite faults in combination operating mechanisms and develops simulation scenarios with varying levels of fault severity. Given the challenges of traditional simulation methods in performing quantitative analysis of core jamming faults and the susceptibility of the core's motion trajectory to external interference, this paper innovatively installs a spring-damping device at the extended core position. This ensures that, during the simulation of core jamming faults, the motion trajectory remains stable and unaffected by external factors, while also enabling precise control over the degree of jamming. As a result, the simulation more accurately reflects real fault conditions, thereby enhancing the accuracy and practicality of diagnostic model outcomes. This study employs the Morlet wavelet transform to convert the current and displacement signals in the time series into time-frequency spectrograms. These spectrograms are then processed using the ResNet50 deep residual neural network for feature extraction and fault classification. The results demonstrate that, when addressing the diagnostic problem of small-sample data for operating mechanism faults, ResNet50, with its residual structure design, exhibits significant advantages. The convolutional layer strategy, which first performs dimensionality reduction followed by dimensionality expansion, combined with the use of the ReLU activation function, contributes to superior performance. This approach achieves a fault recognition accuracy of up to 91.67%.

Keywords: composite faults; fault severity; Morlet wavelet analysis; ResNet50 network

1. Introduction

Circuit breaker operating mechanisms, when subjected to long-term mechanical stress and environmental influences, inevitably exhibit various mechanical damages or defects over time [1]. Operational surveys of high-voltage circuit breaker mechanisms indicate that mechanical defects in the operating mechanism account for 63.8% of the instances of breaker refusal to operate, representing a major contributing factor [2]. The causes of these faults are complex and involve a combination of factors, including the jamming of the opening and closing coil core, significant motion resistance encountered during the core's accelerated movement, as well as physical damage or structural deformation of the coil and related transmission components.

It has been reported that in a 110 kV substation in Liaoning Province, the CD10 model operating mechanism experienced multiple atypical operational incidents within its one-year service period. These incidents were primarily characterized by delayed closing actions and failure of the opening function. Upon thorough investigation, it was confirmed that the fault was caused by a combination of issues, including jamming of the opening and closing coil core, deviation of the actuator rod stroke from the normal range, excessive setting of the limit screw, and twisting deformation of the linkage in the transmission mechanism [3]. Due to the complexity of its mechanical structure, the fault mechanism is often associated with multiple potential failure sources. Therefore, a comprehensive investigation into the failure mechanisms of multi-factor faults in such systems is an indispensable task [4].

Researchers both domestically and internationally have proposed various signal processing methods in previous studies, generally divided into two stages: signal feature extraction and fault classification. Traditional methods, such as modal analysis [5], feature entropy [6], wavelet thresholding [7], and frequency response function estimation [8], have been widely applied in the field of signal feature extraction. Reference [9] applied the Relief algorithm for dimensionality reduction and optimization of the current characteristics of the opening and closing coils, followed by fault classification using a self-organizing map (SOM) network. Reference [10] established characteristic curves for the current waveforms in the opening and closing coils to explore their correlation with the performance of the electromagnetic and mechanical transmission systems. Fault diagnosis and evaluation were precisely carried out through magnetic circuit analysis. Reference [11] integrated coil current signals with additional signal processing techniques to construct a feature vector space and then used support vector machine (SVM) techniques to distinguish between fault types. However, the aforementioned studies mainly focus on identifying single mechanical fault states, and there is a lack of in-depth exploration and systematic research on the dynamic characteristics of operating mechanisms under complex fault conditions, where multiple concurrent or interacting defects are present.

Given the high requirements for data accuracy and completeness in diagnostic models, improper data processing methods inevitably lead to deviations in diagnostic results. To address this issue, researchers both domestically and internationally have increasingly integrated artificial intelligence into equipment fault diagnosis models. Reference [12] extracts the energy distribution features of signal subspaces using wavelet transforms and builds an auto-encoding deep fault diagnosis model through distributed training and labeled fine-tuning. Reference [13] combines the LVQ neural network and DAG-SVM to input zero-point fault current parameters for deep learning, detecting corresponding fault types. However, due to the limited selection of feature parameters and the scarcity of signal resources, the current models face limitations such as high training loss errors and a lack of information correlation. Additionally, methods like radial basis function (RBF) neural networks [14], backpropagation (BP) neural networks [15], support vector machines (SVM) [16], and decision trees [17,18] all belong to the category of traditional shallow learning architectures. Given the instantaneous nature of operating mechanism actions, the complexity of signal transmission paths, and the diversity of external interference factors, these shallow network techniques are often prone to the "curse of dimensionality". This leads to excessively high complexity in the feature space and faces challenges such as the model becoming trapped in local optima and poor generalization performance [19,20].

To address the aforementioned limitations, this study first redefines the classification of faults and their severity levels, designing a series of composite fault scenarios to comprehensively cover the various defect combinations that an operating mechanism may encounter. To accurately capture the dynamic behavior of the operating mechanism under different operating conditions, an efficient current and displacement signal acquisition system platform was developed. Using the Morlet wavelet analysis strategy, time-series signals are transformed into time-frequency images, which serve as feature vectors for subsequent analysis. This approach enhances the temporal and spatial representation capabilities of the features.

The proposed ResNet50-based fault diagnosis method for spring operating mechanisms demonstrates novelty and contributions in the following aspects. Firstly, the deep architecture with residual connections addresses the vanishing gradient problem during deep network training, thereby enhancing the model's learning capability and robustness. Comparative experimental results indicate that ResNet50 outperforms other deep learning models (e.g., MobileNetV2 and Transformer) in handling complex small-sample datasets.

The model effectively addresses the difficulties of training deep networks through its residual connection mechanism, ensuring the efficient propagation of gradients, thereby enhancing the model's performance in handling composite mechanical fault identification tasks. The study uses the CT-26 spring-operated mechanism as the experimental subject, applying the proposed diagnostic model for comprehensive evaluation. The goal is to verify the model's ability and effectiveness in accurately identifying composite mechanical faults in real-world scenarios, providing a scientific foundation and technical support for health management and fault prediction in high-voltage circuit breakers.

2. Simulation Design of Mechanical Faults in the Spring-Operated Mechanism

In this study, we conducted simulated designs for three types of faults—abnormal core gap, coil core sticking, and abnormal core stroke—in the CT-26 spring operating mechanism produced by Taikai High Voltage Switch Co., Ltd. in Shandong, China, under its brand-new condition. The specific design schemes are as follows.

2.1. Design for the Abnormal Core Gap Fault

The core gap distance T quantifies the vertical distance between the top of the coil core (A) and the bottom of the opening lever (B). The specific value of T is precisely controlled by adjusting the nut c. After multiple operations of opening and closing the spring-operated mechanism under dynamic conditions, the accelerated movement of the core collides with the lever, generating an opposing force that causes the positioning nut c to loosen. This loosening effect directly leads to an undesirable increase in the core gap T, thereby increasing the risk of failure when executing the opening command. Based on extensive measurements of core gap values under healthy conditions and discussions with technical production experts, the standard range for T is determined to be between 0.8 mm and 1.2 mm.

As shown in Figure 1, the core gap distance T is controlled by the core adjustment bolt C. Rotating bolt C to the left reduces the gap distance, while rotating it in the opposite direction increases the gap. With repeated opening and closing commands issued to the coil, the opposing surfaces of A and B experience continuous collisions over time. During these impacts, the forces and reaction forces cause bolt C to loosen, leading to an enlargement of the core gap.

Figure 1 illustrates the simulation of the core gap fault. By adjusting the rotation angle of bolt C, the distance T between A and B is made greater than the upper limit of the enterprise's standard, 1.2 mm. The red box in the figure indicates the fault location, which is at the position where the limit lever meets the core. To investigate the operational status of the mechanism under different fault severities, the experiment designs core gap values of 1.6 mm (Level 1 gap anomaly) and 2.0 mm (Level 2 gap anomaly).

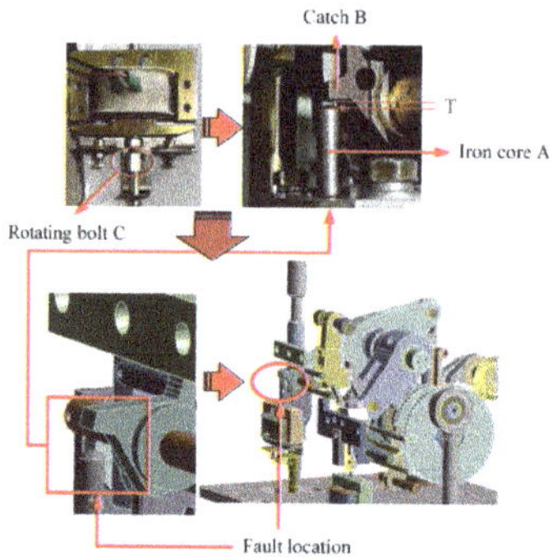

Figure 1. Simulation of core gap fault.

Coil Core A: Refers to the part of the actuating mechanism responsible for driving the trip action of the circuit breaker.

Trip Lever B: A component of the mechanical structure that is responsible for locking and releasing the trip action.

Core Gap T: Refers to the vertical distance from the top of Coil Core A to the bottom of Trip Lever B.

Adjustment Mechanism: (1) The component C nut is a mechanical component used to precisely adjust the core gap distance (T). By rotating or shifting the nut, the distance between the coil core and the trip lever can be fine-tuned to ensure optimal operational performance. (2) The standard range for T was determined to be between 0.8 mm and 1.2 mm. This was based on extensive measurements of core gaps in healthy states, followed by discussions with technical and production experts.

2.2. Design of Core Jamming Fault

The core, as the key component of the operating mechanism's coil, is responsible for receiving and responding to the coil's action commands. When the coil is energized, the electromagnetic attraction acts on the core, causing it to overcome its own resistance and accelerate in motion, ultimately colliding with the upper limit lever to achieve the opening and closing functions of the mechanism. However, mechanical wear over long-term operation can cause the surface of the core to become rough. Additionally, prolonged exposure to a humid environment or lack of anti-rust oil protection can lead to core corrosion. Under the combined effects of these unfavorable factors, the motion resistance of the core increases, necessitating a higher coil current to enhance the electromagnetic attraction in order to ensure the proper operation of the mechanism.

In previous studies, various methods have been used to simulate core jamming, including suspending weights and introducing materials such as foam or wood shavings into the core gap. However, suspending weights can cause excessive oscillation of the core's motion, resulting in unstable force simulation during movement. On the other hand, introducing foam, wood shavings, or similar materials makes it difficult to quantify the degree of jamming, leading to rough data and limited practical engineering guidance.

To address the aforementioned issues, this study introduces the use of a damping device to simulate core jamming. The overall fault simulation design is shown in Figure 2. The original core in the coil is replaced with an elongated core that has threads at the bottom, maintaining the same operational state as the original core. A jamming damping device is installed in the middle of the elongated core. The damping device consists of a damping spring and a rotating bolt Z, which is used to adjust the force of the damping spring. During the upward motion of the core, the damping spring is automatically compressed, successfully simulating the core jamming fault. The force state of the core at this point is shown in Figure 3. The electromagnetic attraction force F_a must overcome both the core's own resistance F_b and the externally applied damping force F_c.

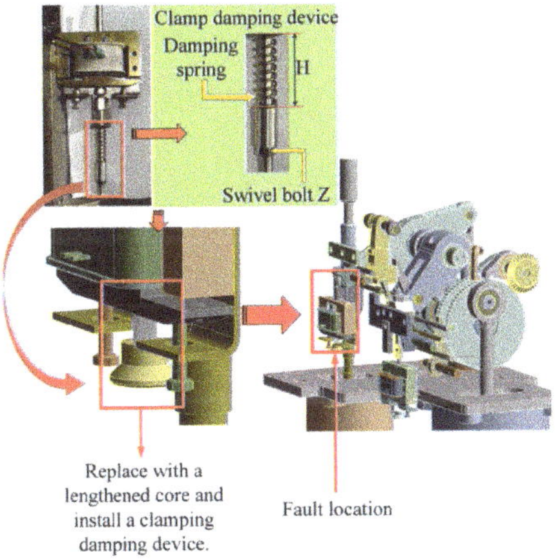

Figure 2. Simulation of iron core jamming fault.

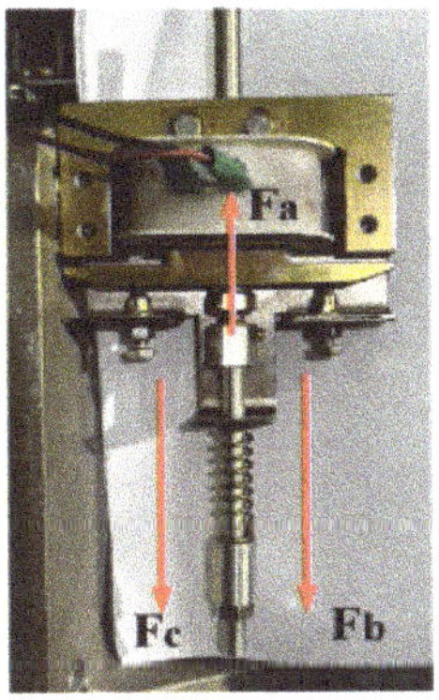

Figure 3. Force state of iron core jamming.

To specifically diagnose the severity of core jamming in the operating mechanism, two different fault severities are set. In Figure 2, the length H of the damping spring represents the severity of the jamming. The unstretched length of the damping spring is 5 cm. In the experiment, the spring length H is set to 4.6 cm (Level 1 core jamming) and 4.2 cm (Level 2 core jamming) for the two fault severity levels.

During the operation of the spring actuation mechanism, the core moves upward or downward during the opening or closing process. In this design, the core compresses the damping spring during its upward movement, simulating the core binding fault. The process is as follows:

Core Moving Upward: When the electromagnetic force acts on the core, the core begins to move upward. During this motion, the damping spring is gradually compressed, generating an opposing resistance.

Force Analysis: During the upward motion of the core, three main forces are at play—(1) electromagnetic attraction, which is the primary force driving the upward motion of the core, (2) the core's own resistance, which includes the core's weight and internal friction, and (3) additional externally applied damping resistance, which is provided by the damping spring, simulating the frictional force encountered when the core binds.

During the upward movement, the core must overcome these resistances to complete the operation. By adjusting the position of the rotating bolt Z, the pre-tension of the damping spring can be altered, thereby regulating the total resistance the core experiences.

2.3. Coil Stroke Fault Design

The overall structure of the opening and closing coil consists of four parts: the fixed structure A, the moving structure B, the coil, and the core. As shown in Figure 4, the distance W between the fixed structure A and the moving structure B is referred to as the core stroke distance. The coil core and structure B form a unified assembly. To maintain the stability of the core's movement and mitigate oscillation and swaying during motion, a damping spring is placed between structures A and B.

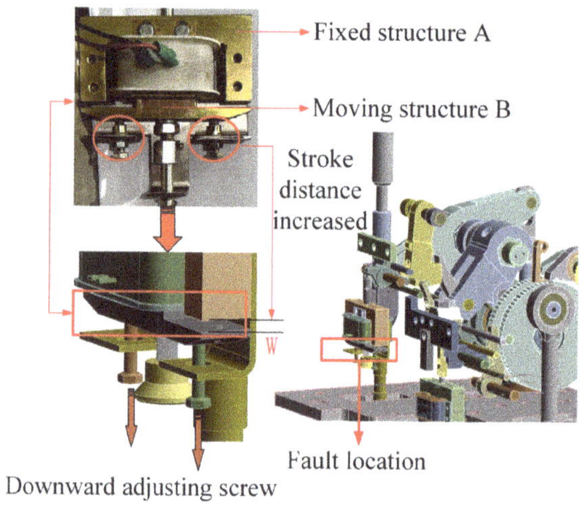

Figure 4. Simulation of abnormal stroke distance fault.

In Figure 4, the constraint bolts at both ends of Structure B, marked by the red circles, are used to precisely adjust the stroke distance W by rotating the bolts. This adjustment mechanism ensures that the actuator remains in optimal working condition during the open/close operations. The technical details are as follows:

(1) Structure and Adjustment Mechanism

Fixed Structure A: This component limits the maximum movement range of the core, ensuring that the core does not exceed the predetermined range under the influence of electromagnetic forces.

Structure B and Constraint Bolts: Structure B consists of two key components—the core and the constraint bolts. By rotating the constraint bolts, their preload force can be adjusted, thus controlling the stroke distance W. The design of the constraint bolts allows for precise control of the core's travel to meet different operational requirements.

(2) Mechanical Analysis During Operation

Electromagnetic Force: When the coil is energized, a strong electromagnetic field is generated, providing an upward electromagnetic force to the core. This force drives the core to overcome gravity and other resistances, completing the open/close operation.

Stroke Distance W: This is defined as the distance from the core's initial position to the maximum displacement point. By adjusting the angle of the constraint bolts, this distance can be precisely controlled, ensuring it stays within the specified range (3.9 mm to 4.3 mm).

(3) Industry Standards and Experimental Setup

According to industry standards, the stroke distance W should be strictly controlled within the range of 3.9 mm to 4.3 mm. To investigate the impact of excessive stroke on the actuator's performance, two abnormal stroke distances were designed for experimental verification:

First-Level Stroke Abnormality: Set the stroke distance W = 4.9 mm to simulate a slight overtravel condition.

Second-Level Stroke Abnormality: Set the stroke distance W = 5.9 mm to simulate a more severe overtravel condition.

3. Design of Composite Faults in the Operating Mechanism and Experimental Data Acquisition System

In the actual operating environment of the mechanism, due to continuous mechanical loads and environmental interactions, the internal components often undergo comprehensive aging and performance degradation. This degradation does not occur in isolation but exhibits the characteristic of multiple simultaneous faults. Previous research has largely been confined to signal feature acquisition and analysis under single fault modes, with isolated exploration of the operational characteristics of the mechanism under each fault condition. This approach overlooks the interactions and superposition effects between faults. Compared to the complex real-world situations, it presents certain simplifications and deviations, failing to adequately reflect the true dynamic behavior and potential fault evolution paths of the operating mechanism when facing multiple simultaneous faults. Therefore, research on fault diagnosis of operating mechanisms urgently requires a more comprehensive approach that is closer to actual working conditions to accurately identify and assess the system's state under multi-fault coexistence.

3.1. Composite Fault Design

This study combines single faults of varying severities to design six types of composite faults. The specific design is shown in Figure 5.

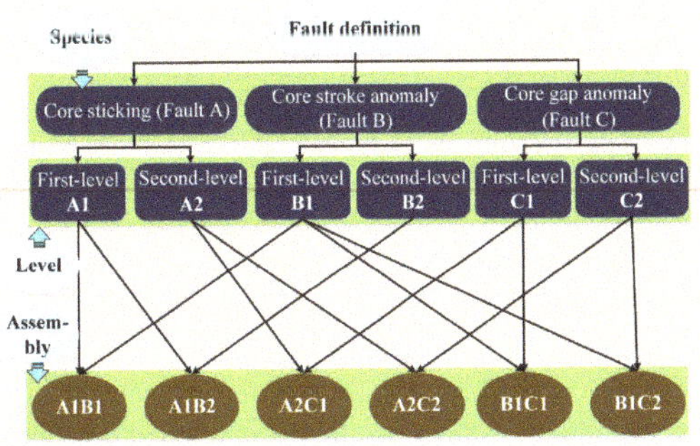

Figure 5. Combination Fault Definition Diagram.

In Section 2 of this paper, the faults related to the core sticking, core stroke anomaly, and core gap anomaly are named as Fault A, Fault B, and Fault C, respectively. Specifically, in the case of core sticking, the first-level sticking (spring length H set to 4.6 cm) is designated as Fault A1, and the second-level sticking (spring length H set to 4.2 cm) is designated as Fault A2. In the case of core stroke anomaly, the first-level anomaly (stroke distance 4.9 cm) is designated as Fault B1, and the second-level anomaly (stroke distance 5.9 cm) is designated as Fault B2. In the case of core gap anomaly, the first-level gap distance (1.6 mm) is designated as Fault C1, and the second-level gap distance (2.0 mm) is designated as Fault C2.

The overall composite faults are categorized into three types, with each category consisting of two levels of severity. These different fault types are selectively combined in pairs of varying degrees. As shown in Figure 5, the composite faults are as follows: first-level core sticking and first-level stroke anomaly (A1B1), first-level core sticking and second-level stroke anomaly (A1B2), second-level core sticking and first-level gap anomaly (A2C1), second-level core sticking and second-level gap anomaly (A2C2), first-level stroke anomaly and first-level gap anomaly (B1C1), and first-level stroke anomaly and second-level gap anomaly (B1C2). These represent a total of six abnormal operating conditions.

3.2. Design of the Composite Fault Acquisition System for the Actuating Mechanism

As shown in Figure 6, the rotational sensor is positioned at the output shaft side of the actuating mechanism. The output end of the sensor is connected to the input terminal of the switch characteristic tester. The tester converts the rotational angle of the output shaft into the displacement distance of the moving contact through an analog-to-digital conversion. The sequential displacement data are then stored in the test record.

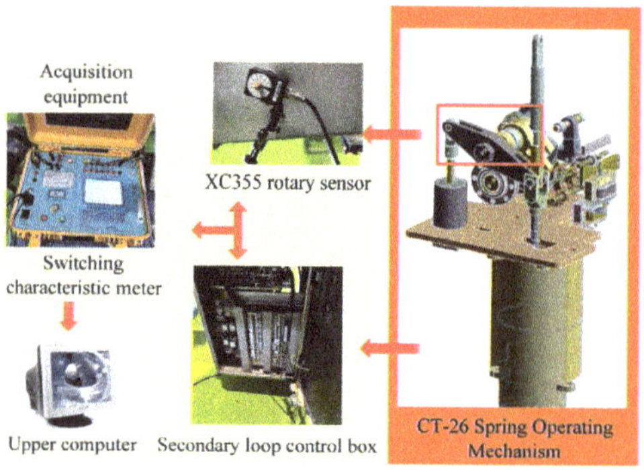

Figure 6. Experimental collection system.

The experimental setup was configured as follows: The breaker's split-off and closing circuits in the actuating mechanism's secondary control loop were connected to the yellow and red ports of the high-voltage switch characteristic tester, with the black port designated as the common terminal. A 220 V power supply was utilized to energize the energy storage circuit. Three-phase contact circuits (A, B, and C) of the circuit breaker were interfaced with the tester's output ports (yellow, green, and red). Voltage regulation was controlled by the characteristic tester, which was configured with a sampling time of 150 ms.

For data acquisition, a circuit breaker switch characteristic tester was employed, featuring six contact status indicators for state verification and wiring validation. The instrument's specifications included a time measurement range of 1–20,000 ms (±(0.005%t + 0.1) ms accuracy), analog input range of 0–5 V, and coil current measurement capability of 0–2/20 A with 10 mA resolution for both DC and AC measurements. Displacement signals from the circuit breaker's dynamic and static contacts were captured using an XC-355 rotary sensor, characterized by 16-bit resolution, ±0.005° accuracy, 9–36 V operating voltage, and 2500 mm effective measuring stroke.

4. Signal Preprocessing and Morlet Wavelet Transform Under Fault Conditions

In this experiment, a detailed fault simulation and data collection were conducted on the opening and closing processes of the actuator using the signal acquisition system shown in Figure 6. The switch characteristic tester was used to control the actuator's operation. Independent experimental trials were carried out for six different types of combined faults, with 50 sets of current and displacement signal samples collected for each fault type. In total, 600 sets of valid data were acquired.

Considering that the raw current signals are inevitably influenced by various external interference factors during the data collection phase, a five-point cubic smoothing technique was applied for data preprocessing to filter out irregular fluctuations and background noise. To further explore the intrinsic features of the signals, wavelet transform analysis was employed. The current and displacement signals, processed through both smoothing and wavelet transform, were then used as input data for the convolutional neural network (CNN) model.

4.1. Five-Point Cubic Smoothing of Current Signals

Figure 7 illustrates the variation of the current signal in the coil of the actuator under the six predefined composite fault modes. The red-marked areas highlight the noise and spikes in the current signal. These disturbances are primarily caused by factors such as magnetic field coupling effects, electromagnetic radiation interference, and leakage coupling, among others. The characteristics of such interference signals are manifested by their high-frequency components and periodic appearance patterns. These features significantly disrupt the time-frequency spectrum distribution during the wavelet transform of the original current signal, reducing the accuracy and reliability of signal analysis.

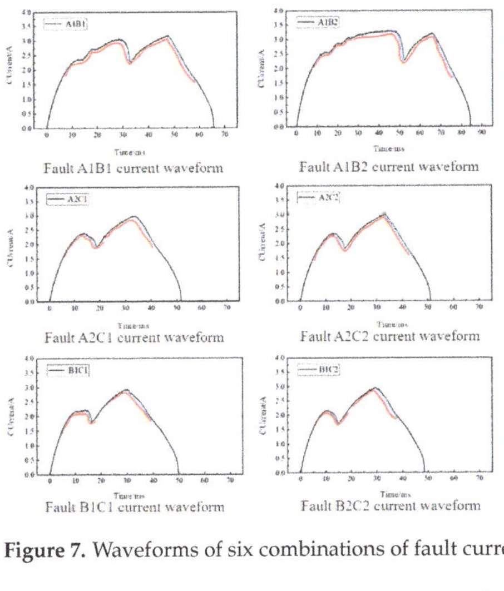

Figure 7. Waveforms of six combinations of fault currents.

This study employs the five-point cubic spline smoothing method, constructing a local cubic polynomial curve with five points to fit the original data. This approach effectively mitigates high-frequency noise interference while preserving the overall shape of the signal. It ensures that the signal is more focused on the true characteristics of the current signal, thereby improving the accuracy and effectiveness of fault diagnosis.

To divide the original current waveform into n evenly spaced points, the current waveform data can be represented as $X_0 < X_1 < X_2 < \ldots < X_{n-1}$. For each interval between two adjacent points, the cubic polynomial is applied, considering the two neighboring points before and after each data point. The cubic polynomial is given as follows:

$$Y = a_0 + a_1 x + a_2 x^2 + a_3 x^3 \tag{1}$$

The coefficients a_0, a_1, a_2, a_3 are determined using the least squares method, and the five-point cubic smoothing formula is obtained as follows:

$$\begin{cases} \overline{Y_{-2}} = \frac{1}{70}(69y_{-2} + 4y_{-1} - 6y_0 + 4y_1 - y_2) \\ \overline{Y_{-1}} = \frac{1}{30}(2y_{-2} + 27y_{-1} + 12y_0 - 8y_1 + 2y_2) \\ \overline{Y_0} = \frac{1}{35}(-3y_{-2} + 12y_{-1} + 17y_0 + 12y_1 - 3y_2) \\ \overline{Y_1} = \frac{1}{35}(2y_{-2} - 8y_{-1} + 12y_0 + 27y_1 + 2y_2) \\ \overline{Y_2} = \frac{1}{70}(-y_{-2} + 4y_{-1} - 6y_0 + 4y_1 + 69y_2) \end{cases} \tag{2}$$

The fundamental principle of the five-point cubic smoothing method is that the number of data points for the curve is greater than five, where $\overline{Y_i}$ represents the improved value of Y_i. Since the sampling data points for the coil current are much greater than five, to ensure symmetry in the smoothing effect, the endpoints of the sampled data are calculated using, $\overline{Y_{-2}}$, $\overline{Y_{-1}}$, $\overline{Y_1}$, and $\overline{Y_2}$, while the midpoint interval points are computed using $\overline{Y_0}$. The denoised and smoothed coil current waveform is shown in Figure 8.

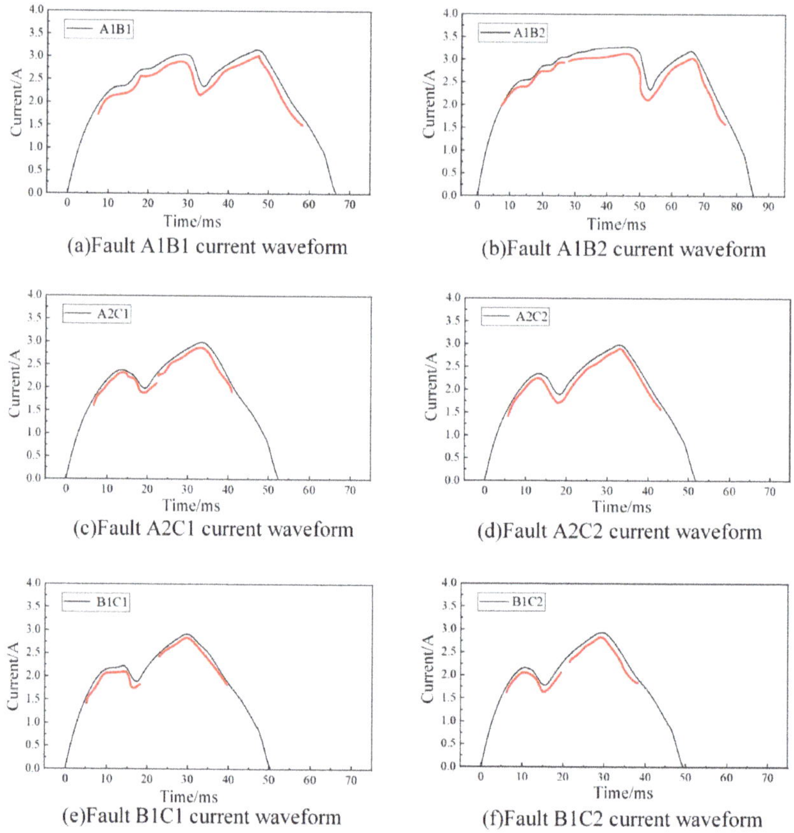

(a)Fault A1B1 current waveform (b)Fault A1B2 current waveform

(c)Fault A2C1 current waveform (d)Fault A2C2 current waveform

(e)Fault B1C1 current waveform (f)Fault B1C2 current waveform

Figure 8. Current waveform after combined fault smoothing.

4.2. Time Domain Waveform of Displacement Signal

Compared to the coil current signal, which may be subject to multiple interferences, these disturbances are significantly suppressed or eliminated in displacement measurements. As a result, the acquired displacement signal exhibits high smoothness and low noise levels, making it suitable for direct frequency-domain transformation analysis. Figure 9 presents the displacement variation curve during the separation operation of the moving contact under six different combination fault scenarios.

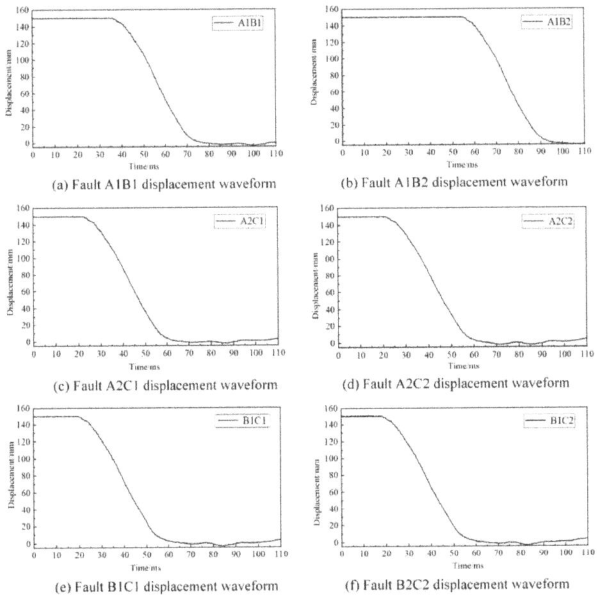

(a) Fault A1B1 displacement waveform (b) Fault A1B2 displacement waveform (c) Fault A2C1 displacement waveform (d) Fault A2C2 displacement waveform (e) Fault B1C1 displacement waveform (f) Fault B2C2 displacement waveform

Figure 9. Combination fault contact displacement waveform.

4.3. Time-Domain Signal Feature Representation via Morlet Wavelet Transform

Traditional Fourier transform, a tool for frequency-domain analysis, decomposes a signal into a linear combination of sinusoidal and cosine basis functions, effectively revealing its spectral composition and providing valuable insights into the signal's frequency structure. However, it assumes that the signal is stationary throughout the observation period, meaning its statistical properties remain constant over time. This assumption limits the applicability of Fourier transform when analyzing non-stationary signals with time-varying characteristics. In particular, Fourier transform struggles with time localization and local feature analysis due to its fixed global time window, which cannot adapt to the dynamic variations of the signal across different time scales.

Wavelet transform, with its time-frequency adaptive characteristics, utilizes basis functions (wavelets) that not only provide multi-resolution properties in the frequency domain but also allow for scaling and translation in the time domain. The wavelet basis can offer finer time resolution in the high-frequency regions (rapidly varying parts of the signal) and broader frequency resolution in the low-frequency regions (slowly varying trends). By dynamically adjusting the width of the analysis window according to the signal's characteristics, it localizes the signal in the time domain while maintaining good resolution in the frequency domain, effectively capturing and analyzing the local time-frequency features of the signal. This overcomes the fixed time window limitation of the traditional Fourier transform, significantly enhancing the local analysis capability of the signal.

To obtain the frequency-domain features of the coil current and contact displacement under different fault conditions, this study employs the Morlet wavelet analysis method to transform the two-dimensional time-domain signals into time-frequency spectrograms. This facilitates the extraction of various signal features from the spectrogram, allowing for a multi-dimensional representation of the signal characteristics of the actuator under multiple combined mechanical faults. The wavelet basis function is a single-frequency complex sinusoidal function with a Gaussian envelope, as shown in Equation (3).

$$\phi(z) = \exp\left(-t^2/2\right) * \exp(j\omega_0 t) \tag{3}$$

The wavelet basis function is obtained by scaling and translating, as shown in Equation (4).

$$\begin{gathered}\phi_{a,b}(z) = \frac{1}{\sqrt{a}} \exp\left[-(t-b)^2/2a^2\right] * \\ \exp[j\omega_0 (t-b)/a]\end{gathered} \tag{4}$$

Within the framework of wavelet analysis, parameter a serves as a scaling factor that adjusts the size of the observation window, directly influencing the degree of stretching or refinement in signal analysis. Increasing the value of a reduces the center frequency of the complex trigonometric function, slows the exponential decay process, and extends the support region of the Gaussian function. The direct consequence of this change is a reduction in the bandwidth of the wavelet in the frequency domain, which enhances the resolution of spectral analysis, allowing for a more detailed and differentiated representation of the signal's time-frequency characteristics. Parameter b controls the translation of the wavelet basis function in the time domain, providing a tool for time localization in signal analysis. This enables feature extraction not only in the frequency domain but also with precise time localization in the time domain.

Figure 10 illustrates the time-frequency representation of the signal after wavelet transformation, providing a clear depiction of how the signal's energy distribution varies with time and frequency. Each time-frequency feature point represents the intensity or activity level of the original signal at a specific moment and frequency, offering valuable time-frequency information. This representation aids in gaining a deeper understanding of the signal structure, identifying potential patterns, and supporting applications such as fault detection.

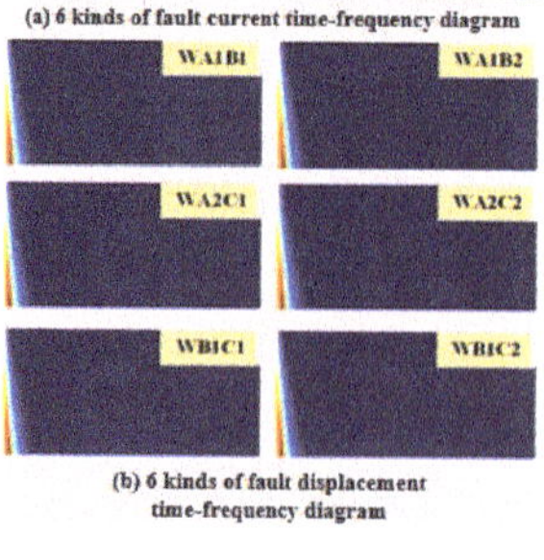

Figure 10. Time frequency map after Moret wavelet transform.

5. Combined Mechanical Fault Diagnosis with ResNet50

In this study, the time-frequency spectrograms of current and displacement under six types of combined faults are used as input data for model training. However, traditional neural networks typically require large datasets for training and model optimization. When the model is overly complex and the training data are limited, overfitting may occur, leading to poor generalization and suboptimal performance on test data. As the network depth increases, the overall performance of the model may deteriorate rather than improve. This study integrates the ResNet50 diagnostic model and compares it with three other models: MobileNetV2, Transformer, and Vision Transformer. The specific technical roadmap is shown in Figure 11.

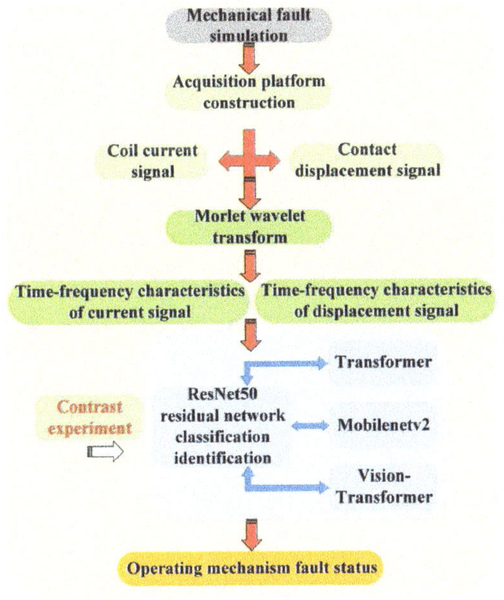

Figure 11. Roadmap of fault diagnosis technology for operating mechanisms.

5.1. Feature Extraction with ResNet50 Model

The ResNet50 deep residual network excels in capturing time-frequency image features from the actuator fault states by applying deep convolution on time-frequency spectrogram datasets. Its core advantage lies in the introduction of residual blocks, which effectively address the vanishing gradient and representation bottleneck issues commonly found in deep neural networks. This design enables the network to learn from the original features at deeper layers, making it highly adaptable to various image recognition tasks. As a result, it significantly enhances the accuracy of actuator image recognition. The overall structure of the model is shown in Figure 12.

To address the issue of insufficient samples, various data augmentation techniques were employed in this study, including random cropping, rotation, and flipping, to generate more diverse training samples. Regarding the dataset imbalance problem, the SMOTE resampling technique and a modified loss function were implemented to ensure optimal model performance on minority classes.

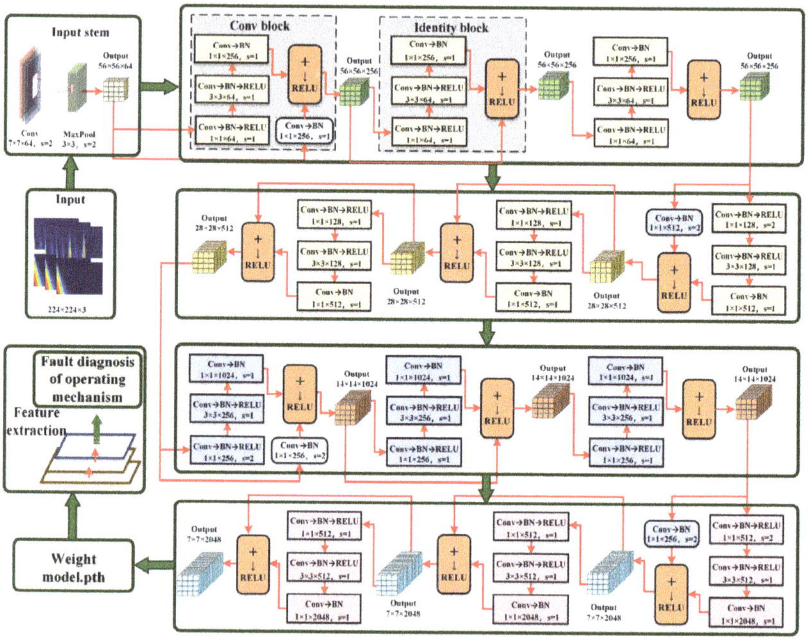

Figure 12. Overall structure of the algorithm in this article.

The overall structure of the ResNet50 algorithm consists of two basic building blocks: the Conv-block and the Identity-block. The Conv-block, also known as the convolutional block, is composed of several convolutional layers designed to extract and recognize key information from the actuator images. The typical structure of a Conv-block includes three parts: the 1 × 1 convolutional layer, which reduces the number of channels in the actuator image, the 3 × 3 convolutional layer for feature extraction, and another 1 × 1 convolutional layer to maintain dimensionality and restore the original number of channels. Conv-blocks are typically placed at the input stage to learn low-level features from the time-frequency images of the actuator.

The Identity-block, based on identity mapping, is also composed of three convolutional layers, and its structure is similar to that of the Conv-block. The 1 × 1 convolutional layer is used to maintain or reduce the number of channels, while the 3 × 3 convolutional layer is used for feature extraction. The final 1 × 1 convolutional layer restores the number of channels. The Identity-block maintains consistent input and output dimensions, and the feature information from the input is directly added to the output through a skip connection. This mechanism effectively addresses the issues of vanishing and exploding gradients typically encountered in deeper networks.

5.2. Model Fault Diagnosis Results and Comparative Experiments

5.2.1. Model Development Environment and Dataset Preparation

The model was built using the PyTorch 1.8.0 framework. In the experiments, the number of training epochs was set to 200, weight decay was set to 0.0005, and the learning rate was adjusted within the range of 0.01 to 0.0001. The stochastic gradient descent (SGD) optimizer was employed. The dataset was divided into two categories: current and displacement datasets. For each of the six fault combinations, the current dataset (denoted by the symbol 'D') and the displacement dataset (denoted by the symbol 'W') were created.

Each dataset was split in a 4:1 ratio, with 40 samples used for training and 10 samples for testing, resulting in a total of 600 images across both datasets.

5.2.2. Comparison of Fault Diagnosis Model Performance for Actuating Mechanism

This study compares the performance of the ResNet50 model in diagnosing small-sample faults in the spring-actuating mechanism with three other models: MobileNetV2, Transformer, and Vision Transformer. The loss function for each model is shown in Figure 13.

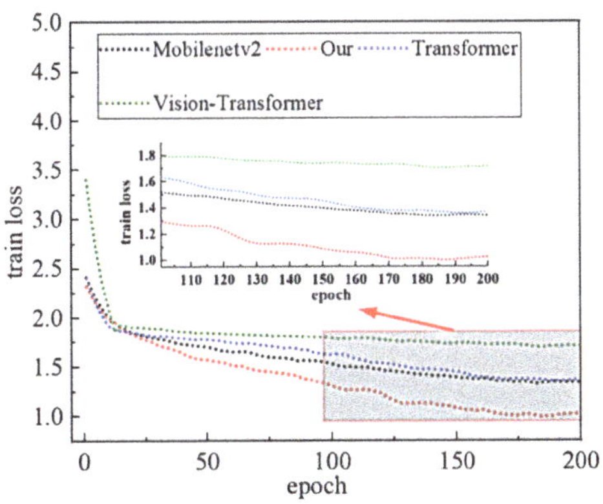

Figure 13. Loss function value transformation curve.

The loss function provides a quantitative method to evaluate the discrepancy between the model's predictions and the actual observed values. By minimizing the value of the loss function, the model parameters are optimized to improve prediction accuracy, prevent overfitting, and introduce regularization terms to balance model complexity and predictive capability.

As shown in Figure 13, after the 100th training epoch, the training loss of the V-T diagnostic model converges at 1.8. The Transformer and MobileNetV2 models exhibit better training loss, converging at 1.4. All three models have a loss greater than 1. However, when applied to small-sample data classification, ResNet50 shows a significant improvement, with the loss converging to 0.85 after the 170th epoch, demonstrating a notable enhancement in model learning performance.

In this study, model performance is evaluated using precision and recall metrics. Precision measures the proportion of true positive instances among all instances predicted as positive by the classifier or retrieval system. Recall, on the other hand, focuses on the proportion of actual positive instances correctly identified by the classifier or system, assessing its ability to detect all true positives. A higher recall indicates that the system can effectively identify all actual positive instances. A comparison of the evaluation metrics is shown in Figure 14.

As shown in Figure 14, the ResNet50 algorithm proposed in this study significantly outperforms the other three algorithms. The improved performance of this model can be attributed to its deeper network architecture, which enables the learning of more complex features, along with the introduction of residual connections. These connections address the gradient vanishing problem during the training of deep networks. Each residual block contains two convolutional layers and a skip connection, which not only ensures efficient gradient propagation but also enhances the model's training efficiency and performance. The model achieved precision and recall rates of 91.8% and 91.67%, respectively.

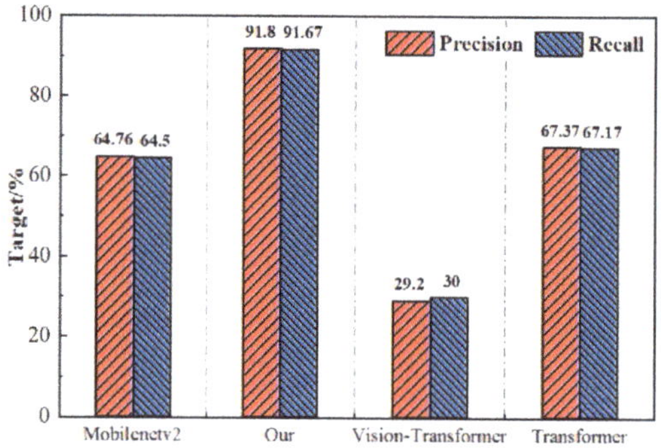

Figure 14. Comparison of P-R Index Performance.

To provide a more intuitive comparison of how different algorithms perform on small-sample fault diagnosis data from the actuator mechanism, confusion matrices for all four algorithms were plotted. These matrices directly reflect the accuracy of each algorithm in recognizing signals from different fault combinations. The confusion matrix comparison for the four models is shown in Figure 15, where the horizontal axis represents the predicted labels, and the vertical axis represents the actual labels.

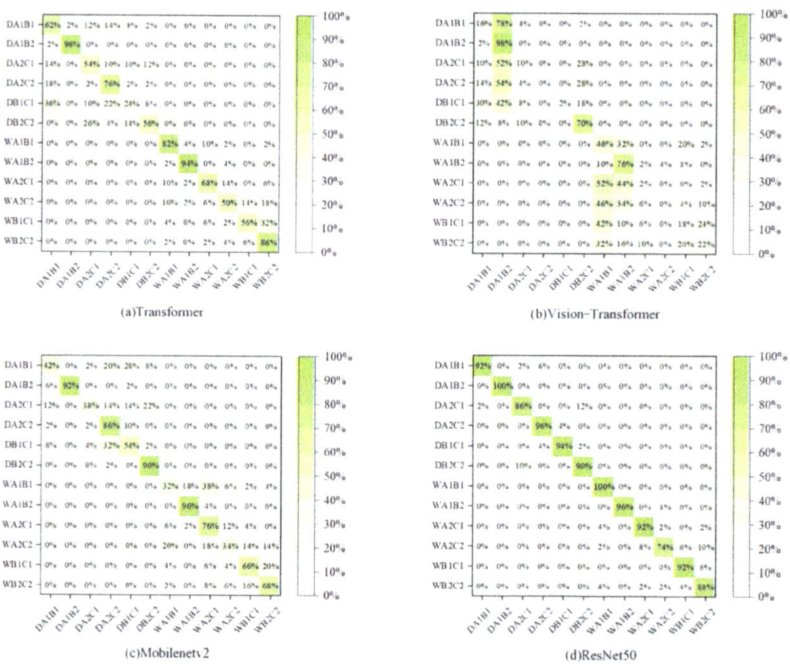

Figure 15. Comparison of model fault diagnosis evaluation performance.

Figure 15a shows the fault diagnosis results for the Transformer model. The overall performance of this model is moderate, with accuracy rates for identifying the actuator combination fault images DA1B2, WA1B1, WA1B2, and WB2C2 being 98%, 82%, 94%, and 86%, respectively. However, the recognition rates for the remaining fault images are all below 80%.

Figure 15b shows the fault diagnosis results for the Vision Transformer model, where only the DA1B2 recognition rate reaches 98%, and the performance for other faults is poor. Among the four models, the best performance is seen in the (b) MobileNetV2 model, which also incorporates residual networks and uses depthwise separable convolutions with linear activation during dimensionality reduction. This model achieves an accuracy of over 90% for identifying DA1B2, DB2C2, and WA1B2 images, with a recognition rate of 86% for DA2C2. The (d) ResNet50 model used in this study performs the best overall. By introducing residual networks and employing standard convolutions with ReLU activation functions, and using nonlinear activation during both upsampling and downsampling, the model achieves an overall fault recognition accuracy of 91.67%. The recognition rate for the WA2C2 fault image is 74%, while all other fault images exceed 85%, with DA1B2 and WA1B1 fault images achieving perfect accuracy of 100%.

Top-1 Accuracy refers to whether the model's predicted highest probability class matches the true class. If the model's predicted highest probability class corresponds to the true class, the prediction is considered correct; otherwise, it is considered incorrect. Top-1 Accuracy represents the proportion of correct predictions made by the model. On the other hand, Top-5 Accuracy indicates whether the true class is within the top five predicted classes with the highest probabilities. If the true class is present in the top five predictions, the prediction is deemed correct; otherwise, it is incorrect.

As shown in Table 1, the ResNet50 model achieves the highest Top-1 Accuracy at 91.67%, followed by the Transformer model with 67.17%. The accuracies for MobileNetV2 and Vision Transformer are relatively lower, at 64.50% and 30.00%, respectively.

Table 1. Diagnostic model comparison parameters.

Modules	Image Size	Top-1 Accuracy	Top-5 Accuracy	Parameters	GFLOPs
MobileNetV2	[224 224]	64.50%	99.67%	3.505 M	0.654 G
Swin Transformer	[224 224]	67.17%	99.33%	87.705 M	30.340 G
Vision Transformer	[224 224]	30.00%	95.33%	86.416 M	33.727 G
ResNet50	[224 224]	91.67%	99.67%	25.557 M	8.267 G

In terms of parameter count (Params) and computational complexity (GFLOPs), the Swin Transformer and Vision Transformer models exhibit higher complexity, with parameter counts of 87.705 M and 86.416 M, and computational complexities of 30.340 G and 33.727 G, respectively. In contrast, the ResNet50 and MobileNetV2 models have lower complexity, with parameter counts of 25.557 M and 3.505 M, and computational complexities of 8.267 G and 0.654 G, respectively.

The ResNet50 fault diagnosis model demonstrates optimal performance in terms of accuracy, while maintaining relatively low complexity, thus achieving a balance between efficiency and performance. In contrast, although the Transformer and Vision Transformer models exhibit significantly increased complexity, they do not surpass ResNet50 in predictive accuracy, indicating that the higher resource demands do not directly translate into better performance. The MobileNetV2 model performs mediocrely in both aspects, failing to achieve top-tier accuracy, yet maintaining a lower complexity.

6. Conclusions

This paper addresses the complex mechanical failure issues faced by the CT-26 spring actuating mechanism in real-world operational environments. An innovative fault simulation strategy was designed to detect the combination of faults in the actuating mechanism, utilizing wavelet transform and deep residual networks.

(1) Due to the challenges in quantitative analysis and the inadequate simulation effects in traditional core jamming fault simulations, one of the key innovations of this paper is the design and successful integration of a novel spring damping device at the core position. The ingenuity of this device lies in its ability to directly simulate the core jamming state, ensuring high fidelity and controllability during the simulation process. This significantly enhances the accuracy and practicality of fault simulations, providing an effective solution to the long-standing bottlenecks in simulation technology.

(2) In this study, six types of combined faults were designed. During the fault signal preprocessing phase, the Morlet wavelet transform technique was applied to convert the current and displacement signals in the combined fault time series into images rich in time-frequency information.

(3) The ResNet50 deep residual network was employed, leveraging its unique Conv-block and Identity-block modules to address the issues of gradient vanishing and explosion in the diagnosis of time-frequency data for operating mechanisms. The model's recognition performance was enhanced through specialized downsampling and upsampling convolution strategies. An average accuracy of 91.67% was achieved in identifying twelve types of fault signals. With a parameter count of 25.557 million and a complexity of 8.267 gigaflops, computational resources and time were conserved without compromising diagnostic accuracy. This underscores the model's high precision and efficiency in the fault diagnosis of high-voltage circuit breakers.

Author Contributions: Conceptualization, H.S.; formal analysis, H.S., Y.J., J.Z., X.L., M.Z. and M.Y.; Investigation, H.S., Y.J., J.Z., X.L., M.Z., M.Y. and X.W.; writing—original draft, X.W. and H.Y.; writing—review and editing, H.S. and H.Y. All authors have read and agreed to the published version of the manuscript.

Funding: This work was supported by the Innovation Project of China Southern Power Grid Co., Ltd. (Grant No. YNKJXM20222318 and No. YNKJXM20222388).

Data Availability Statement: Data are contained within the article.

Conflicts of Interest: Author Yang Mingkun was employed by the company Electric Power Research Institute of Yunnan Power Grid Co., Ltd. The remaining authors declare that the research was conducted in the absence of any commercial or financial relationships that could be construed as a potential conflict of interest.

References

1. Popov, I.; Jenner, D.; Todeschini, G.; Igic, P. Use of the DMAIC Approach to Identify Root Cause of Circuit Breaker Failure. In Proceedings of the 2018 International Symposium on Power Electronics, Electrical Drives, Automation and Motion (SPEEDAM), Amalfi, Italy, 20–22 June 2018; pp. 996–1001.
2. Yang, Q.; Ruan, J.; Zhuang, Z.; Huang, D. Chaotic analysis and feature extraction of vibration signals from power circuit breakers. *IEEE Trans. Power Deliv.* **2020**, *35*, 1124–1135. [CrossRef]
3. Wang, J.; Zhang, X.; Sun, G.; Hu, W. Failure cause and handling of CD10 type electromagnetic operating mechanism. *Rural. Electrif.* **2005**, *11*, 46–48.
4. Ma, Q.; Rong, M.Z.; Jia, S.L. Study of switching synchronization of high voltage breakers based on the wavelet packets extraction algorithm and short time analysis method. *Proc. CSEE* **2005**, *25*, 149–154.
5. Sun, Y.H.; Wu, J.W.; Lian, S.J.; Zhang, L.M. Extraction of vibration signal feature vector of circuit breaker based on empirical mode decomposition amount of energy. *Trans. China Electrotech. Soc.* **2014**, *29*, 228–236.
6. Sun, L.; Hu, X.; Ji, Y. Fault diagnosis or HV circuit breakers with characteristic entropy of wavelet packet. *Autom. Electr. Power Syst.* **2006**, *30*, 62–65. (In Chinese)
7. Zhang, L.; Huang, X.; Jiang, B.; Hu, H.; Hu, X. On line monitoring technology of mechanical characteristics of circuit breaker based on contact stroke measurement. *J. Xi'an Polytech. Univ.* **2018**, *32*, 302–310.

8. Razi-Kazemi, A.A.; Vakilian, M.; Niayesh, K.; Lehtonen, M. Circuit-Breaker Automated Failure Tracking Based on Coil Current Signature. *IEEE Trans. Power Deliv.* **2014**, *29*, 283–290. [CrossRef]
9. Zhang, H.; Zhao, L.H.; Jing, W.; Yang, W.; Jiang, H.Y.; Zhu, L.L.; Rong, Q.; Fu, R.R. Condition assessment of the circuit breaker operating mechanism based on relief feature vector optimization and SOM network. *High Volt. Appar.* **2017**, *53*, 240–246. (In Chinese)
10. Zhao, L.; Zhang, H.; Jing, W. Research onoperating mechanism performance of high voltage circuitbreaker based on thecoil current. *J. Sichuan Univ.* **2015**, *47*, 146–152.
11. Chen, C.; Li, X.R. Fault diagnosis method of circuitbreaker operating mechanism based on coil currentanalysis. *J. Zhejiang Univ.* **2016**, *50*, 527–535.
12. Chen, X.; Feng, D.; Lin, S. Mechanical Fault Diagnosis Method of High Voltage Circuit Breaker Operating Mechanism Based on Deep Auto-encoder Network. *High Volt. Eng.* **2020**, *46*, 3080–3088.
13. Ma, H.; Xu, Y.; Wei, H.; Liu, Y. Mechanical fault diagnosis method of high voltage circuit breaker based on LVQ neural network. *High Volt. Appar.* **2019**, *55*, 30–36. (In Chinese)
14. Hu, X.; Qi, M.; Ji, Y.; Yu, W. On line monitoring and fault diagnosis of high voltage circuit breakers based on radial basis function networks. *Power Syst. Technol.* **2001**, *25*, 41–44.
15. Wei, W.; Shi, Y. The application of improved PSO-BP algorithm in nonlinear function approximating. *Microelectron. Comput.* **2017**, *34*, 112–115.
16. Hu, Q.; Wang, R.; Zhan, Y. Fault diagnosis technology based on SVM in power electronics circuit. *Proc. CSEE* **2008**, *28*, 107–111. (In Chinese)
17. Sun, Z.; Sun, Z.; Wei, J. Research on power transformer fault diagnosis hased on decision tree support vector machine algorithm. *J. Electr. Eng.* **2019**, *14*, 52–55.
18. Wang, F. *The Application of Decision Tree Algorithm on Fault Diagnosis System for Mechanical Equipment*; Huazhong University of Science and Technology: Wuhan, China, 2013.
19. Zhou, J.; Zhao, H.; Lin, S.; Si, H.; Li, B. Fault diagnosis of high-voltage circuit breaker based on optimized random forest algorithm. *Electron. Meas. Technol.* **2018**, *41*, 95–98.
20. Ma, S.; Chen, M.; Wu, J.; Wang, Y.; Jia, B.; Jiang, Y. High-voltage circuit breaker faultdiagnosis using a hybrid feature transformation approach based on random forest and stacked auto-encoder. *IEEE Trans. Ind. Electron.* **2018**, *66*, 9777–9788. [CrossRef]

energies

MDPI

Article

Research on Diagnostic Methods for Zero-Value Insulators in 110 kV Transmission Lines Based on Spatial Distribution Characteristics of Electric Fields

Lei Zheng [1,2], Pengxiang Yin [1,2,*], Jian Li [1,2], Hui Liu [3], Tao Li [1,2] and Hao Luo [1,2]

[1] NARI Group Corporation Ltd., Nanjing 211106, China; zl202501@126.com (L.Z.); li01250109@yeah.net (J.L.); litao202412@163.com (T.L.); luohao138@yeah.net (H.L.)
[2] Wuhan NARI Limited Liability Company, State Grid Electric Power Research Institute, Wuhan 430206, China
[3] Grid Shandong Electric Power Research Institute, Jinan 250003, China; lh02192025@126.com
* Correspondence: yinpx2580@163.com

Abstract: Porcelain insulators in power systems are subject to prolonged mechanical and electrical loads, as well as environmental factors such as climate variations. These conditions often lead to degradation of insulation performance and structural damage, resulting in a decrease in insulation resistance and the formation of cracks, which in turn produce "zero-value" insulators. The presence of zero-value insulators significantly increases the risk of pollution flashovers and electrical arcing, with flashover occurrences possible even under normal operating voltages. This poses a severe threat to the safe and stable operation of the power grid. This study develops a high-fidelity simulation model of insulator strings containing zero-value defects for a 110 kV transmission line. The impact of variations in the position and quantity of zero-value insulators on the spatial electric field distribution is analyzed in detail. Based on the electric field changes, a detection method for zero-value insulators is proposed. Additionally, a prediction model for the electric field strength of insulators with zero-value defects is developed using a Multilayer Perceptron (MLP) neural network. A spatial electric field distribution database for insulator strings containing zero-value defects is also established. The accuracy of the model is validated through laboratory testing.

Keywords: zero-value insulators; electric field distortion; finite element simulation; electric field distribution; MLP neural network algorithm

1. Introduction

The insulator, as a core component of overhead transmission lines, is essential for the stable and safe operation of the power grid [1,2]. Porcelain insulators, known for their superior insulating properties, are susceptible to internal degradation when exposed over time to complex environmental conditions and high voltages. This degradation, affecting the adhesive within the porcelain, leads to a continuous decrease in internal resistance, eventually resulting in zero-value insulators [3,4]. In overhead lines, the presence of zero-value insulators significantly increases the risk of insulator string breakage or detachment, posing a serious threat to grid stability. Therefore, monitoring the condition of insulators is of critical importance to ensure grid safety [5–7].

Traditional insulator inspection methods, such as insulation resistance testing and spark gap testing, involve high-risk, contact-based procedures requiring personnel to ascend transmission towers. These methods are labor-intensive and time-consuming, and the insulation resistance method requires a power outage [8]. In light of these challenges, researchers have explored non-contact methods, including ultraviolet and infrared imaging. However, these techniques are often vulnerable to environmental interference, leading to potential issues of missed or false detections [9–11]. To enhance the efficiency of zero-value detection, the electric field distribution method has emerged as a promising alternative. This approach leverages the spatial electric field characteristics of insulators to enable non-contact inspection. It offers greater immunity to environmental factors, lower detection costs, and yields relatively accurate measurements.

Research has been conducted on insulator degradation detection based on the electric field distribution method. M. Ramesh et al. used an electro-optic probe based on the Pockels effect to measure electric fields, demonstrating differences between defective and healthy insulators [12]. Yangchun Cheng et al. proposed a novel live detection method for zero-value insulators and verified experimentally that the location of zero-value insulators could be identified by changes in the electric field distribution curve [13]. Yi Xu et al. developed a detection device based on electric field distribution principles; however, the device requires scanning the entire insulator string's field distribution curve, resulting in low detection efficiency [14]. To improve detection efficiency, Dongdong Zhang et al. introduced an innovative focus on local electric field detection, demonstrating its effectiveness in identifying zero-value insulators through simulation, opening a new pathway for improving detection efficiency in transmission lines [15].

Currently, the majority of research is confined to purely simulation analyses, lacking essential experimental validation, and thus fails to offer definitive guidance on the optimal detection distance in field inspections. Furthermore, there exists a significant data deficiency concerning the spatial electric field distribution characteristics of porcelain insulator strings. To address these issues, this study employed finite element simulation techniques to investigate the detection of degraded insulators. A testing platform for degraded insulator detection was developed, and the effects of detection distance, the position of zero-value insulators, and their quantity on the electric field distribution of insulator strings were systematically analyzed. The distribution patterns of electric field intensity under various zero-defect conditions in insulator strings were subsequently revealed. Furthermore, a multi-layer perceptron (MLP) neural network model was utilized to construct a spatial electric field distribution dataset for insulator strings with zero-value defects, leading to the establishment of a corresponding database. The findings of this study provide robust theoretical and data support for field detection of zero-value insulators based on the electric field method.

2. Simulation Model of Suspension Insulator Strings

In this section, a 1:1 three-dimensional electric field simulation model of a tower-insulator string was established using Comsol 6.2 finite element simulation software. Based on electrostatic principles, a method for calculating the electric field distribution within the insulator string and its surrounding space was proposed, along with boundary conditions for the entire electric field simulation model.

2.1. Model Development

The U70BP-160 porcelain insulator (Dalian Insulator Group T&D Co., Ltd. Dalian, China) was selected as the study object, and a 3D simulation model was created in Comsol, as shown in Figure 1b. The model includes key components: the ceramic steel feet, steel cap, and cement adhesive. The porcelain insulator has a height of 146 mm and a diameter of 280 mm, with the steel cap's maximum diameter being 100 mm.

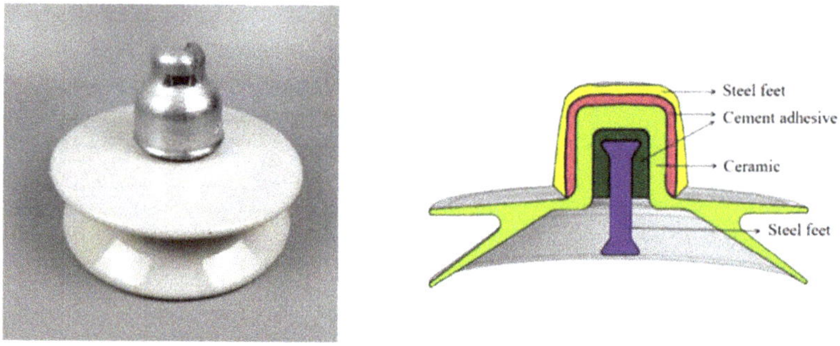

(a) (b)

Figure 1. Porcelain insulator: (**a**) Physical; (**b**) Model.

For zero-value insulators, the primary areas of degradation are at the connections among the steel cap, porcelain body, and steel feet. In this model, through-connections between the steel cap, porcelain, and cement adhesive were created, and the entire through-region was set as a conductive metal material to simulate a zero-value defect. The relative permittivity of the defect area is configured to be at least 1000 times that of the surrounding non-metal conductors.

As shown in Figure 2, the model includes the tower, insulator string, upper and lower fittings, conductor, and an air domain surrounding the insulator string. The electrical configuration comprised a seven-unit U70BP-160 porcelain insulator string. A 20-m conductor assembly with surrounding air domains (radially extending 5 m from the insulator string) was computationally modeled to characterize the spatial electric field distribution. Figure 3 shows the porcelain insulator string model, composed of seven identical insulators numbered from 1 to 7 from bottom to top, where Insulator 1 is connected to the conductor and Insulator 7 is connected to the tower. The parameter d represents the distance from various measurement points to the outer edge of the insulator string, with the electric field detection path along the straight line L, defined as the axial direction in space. Internal component dimensions and material properties were specified, and each part of the insulator was populated with corresponding materials.

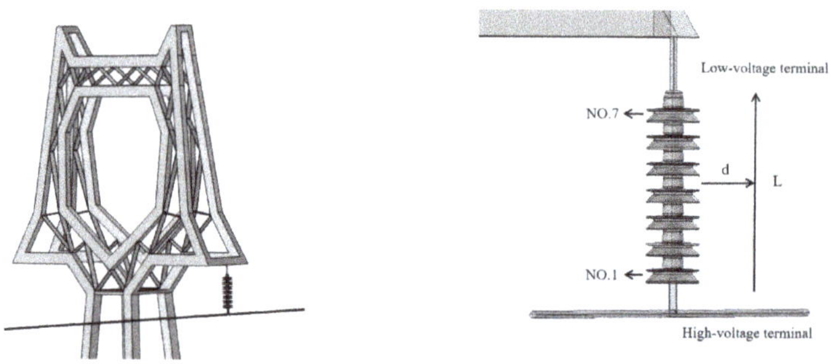

(a) (b)

Figure 2. Simulation model of porcelain insulator string: (**a**) Insulator String; (**b**) Detection path.

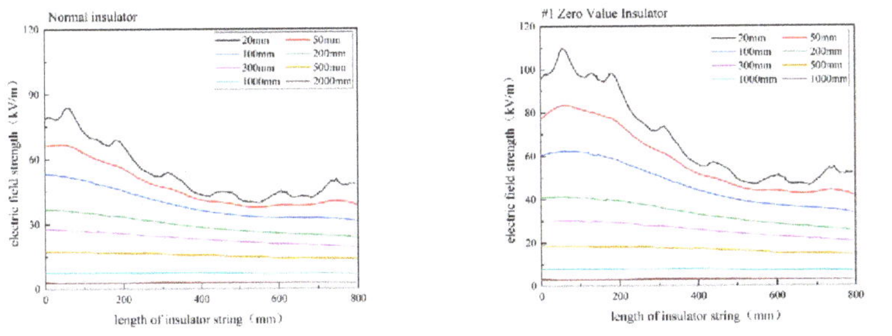

(**a**) (**b**)

Figure 3. Electric field distribution at different detection distances: (**a**) Normal insulator; (**b**) #1 Zero Value Insulator.

An electrostatic analysis approach was applied for simulation calculations, with primary material parameters outlined in Table 1. The electric field simulation employs electrostatic field boundary conditions, where an AC phase voltage of 63.5 kV is applied to the high-voltage end of the insulator string (conductor, connecting hardware, and steel foot of the first insulator). The tower, the steel cap of the last insulator, and the connecting hardware are grounded, with the air domain enforced at 0 potential.

Table 1. Key material parameters for electric field calculation.

Materials	Relative Permittivity
Steel Cap, Steel Foot	1×10^8
Cement Binder	14
Ceramic Components	6
Conductor	1000
Pole Tower	2000
Fittings	1000

2.2. Finite Element Calculation Method and Boundary Condition Setup

In the three-dimensional electrostatic field model presented in this paper [16], the potentials of the tower, conductor, and insulator satisfy Poisson's equation based on the principles of electrostatics, as shown in Equation (1).

$$\nabla^2\varphi = \frac{\partial^2\varphi}{\partial x^2} + \frac{\partial^2\varphi}{\partial y^2} + \frac{\partial^2\varphi}{\partial z^2} = -\frac{\rho}{\varepsilon} \quad (1)$$

In these equations, ∇^2 denotes the Laplace operator, ε represents the permittivity of the dielectric medium, ρ is the spatial charge density, and φ is the electric potential.

To obtain the electric field distribution characteristics of a ceramic insulator string, it is necessary to establish an electric field control equation specific to the insulator string. A 63.5 kV AC phase voltage is applied to the high-voltage end of the insulator string, and the electric field within and around the surrounding air is analyzed. Assuming the influence of space charges is negligible, the solution can be derived using the electrostatic field method. In the simulation model presented, the influence of space charges on the electric field simulation is neglected. Consequently, the resulting potential function Φ satisfies the following equation:

$$\nabla^2\varphi = 0 \quad (2)$$

In the simulation calculations for the test model, first-type boundary conditions are uniformly applied as follows:

$$\varphi|_{\Gamma_0} = 0 \tag{3}$$

$$\varphi|_{\Gamma_k} = U_0 \tag{4}$$

In these equations, Γ_0 represents the boundary of the air region and the ground, while Γ_k denotes the high-potential simulation model. Equations (2) through (4) collectively define the boundary conditions for the entire electric field simulation model.

3. Spatial Electric Field Distribution Characteristics of Insulator Strings Under Various Conditions

This section investigates the impact of detection distance (20 mm to 2000 mm) on the electric field distribution of a 110 kV insulator string through simulation. The effects of the position and number of zero-value insulators under different conditions (single insulators, double insulators with discontinuous zero-value defects, and double insulators with continuous zero-value defects) on the electric field intensity are also analyzed. A comparison between the zero-value defects and normal conditions reveals the specific influence of detection distance and zero-value conditions on the electric field distribution.

3.1. Effects of Detection Distance on the Spatial Electric Field of Insulator Strings

In 110 kV insulator strings, the electric potential decreases uniformly from the high-voltage end to the low-voltage end, with field intensity highest near the high-voltage end, diminishing in the middle, and slightly increasing near the low-voltage end. To investigate the effects of detection distance and zero-value defects on the spatial electric field distribution, the first insulator was set as a zero-value insulator and spatial electric field distribution simulations were conducted across a range of detection distances from 20 mm to 2000 mm. The resulting electric field intensity distribution curves are shown in Figure 3.

Analysis of Figure 3a reveals that, under normal conditions, the electric field strength increases as the detection point moves closer to the insulator string, exhibiting a "U" shaped distribution. At d = 20 mm, the field strength fluctuates due to the insulator's shape, while at d = 2000 mm, it becomes nearly linear. Figure 3b indicates that the #1 zero-value insulator causes significant distortion, particularly at 100 mm, with the distortion effect decreasing along the string. The comparison reveals that the zero-value effect is most pronounced at 20 mm, with the response diminishing beyond 300 mm. Therefore, the detection distance should be maintained between 20 mm and 300 mm to accurately capture the electric field changes.

3.2. Effects of Zero-Value Insulator Position on the Spatial Electric Field of Insulator Strings

To investigate the effect of insulator degradation location on the surrounding field strength, simulations were conducted by varying the relative permittivity to model conditions with single, discontinuous dual, and continuous dual zero-value insulators. The electric field distribution of the insulator string was analyzed under each condition to obtain spatial electric field characteristic curves, allowing for a comparison of how different degradation locations impact field strength variation.

3.2.1. Electric Field Distribution Characteristics with a Single Zero-Value Insulator

In the single insulator zero-value scenario, where the second insulator is set as zero-value, the electric field distribution at d = 20 mm and d = 200 mm is compared, as shown in Figure 4. Under normal conditions, the electric field transitions from a fluctuating shape to a smoother "U" shape as the distance increases. With the #2 zero-value insulator, the electric field is significantly distorted at that location, with a maximum distortion rate of

30% at d = 20 mm. The field decreases below the zero-value insulator and increases above it, with a smooth transition at the zero-value point, demonstrating the significant impact of the zero-value insulator on the surrounding electric field.

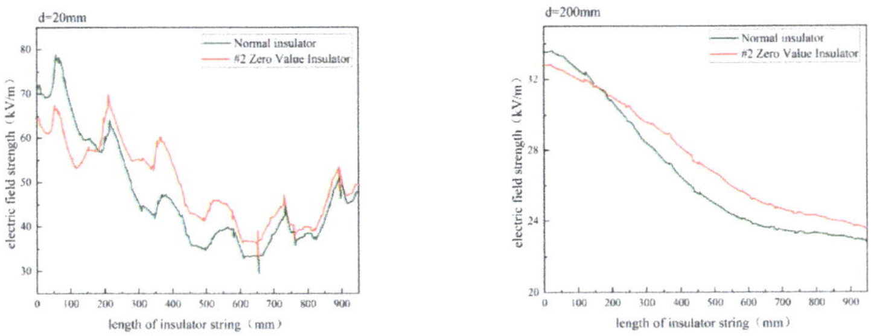

(a) (b)

Figure 4. Electric field distribution under #2 Zero-value condition: (**a**) d = 20 mm; (**b**) d = 200 mm.

To further analyze the impact of the location of a single degraded insulator on the electric field distribution, simulations were conducted with insulators 1 through 7 set as zero-value insulators. The electric field distributions at d = 20 mm and d = 200 mm were then compared. Figure 5 compares the electric field variations under normal conditions and with zero-value insulators at different positions. Under normal conditions, the electric field decreases from the high-voltage end (No. 1) to the low-voltage end (No. 7) before increasing again. The zero-value insulators significantly affect the field distribution, especially at the high-voltage end (No. 1), where the peak electric field reaches 109.6 kV/m and the distortion rate is 50% at d = 20 mm. This study shows that the location of insulator degradation, particularly at the high-voltage and tower ends, significantly influences the electric field distribution.

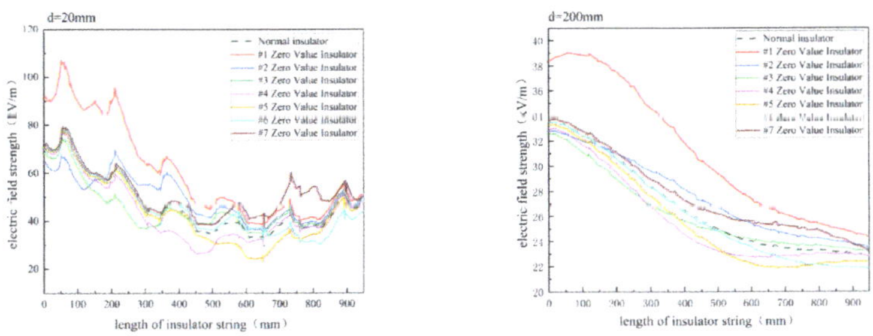

(a) (b)

Figure 5. Electric field distribution under different single insulator zero-value conditions: (**a**) d = 20 mm; (**b**) d = 200 mm.

3.2.2. Electric Field Distribution Characteristics with Discontinuous Double Zero-Value Insulators

Building on the analysis of a single zero-value condition, the effect of discontinuous dual zero-value insulators (#2 and #4) on the electric field distribution outside the insulator string was further examined, as shown in Figure 6. When Insulators #2 and #4 were set as zero-value, the electric field distribution showed significant changes at both d = 20 mm and d = 200 mm. Each degraded insulator had an effect similar to that of a single zero-value unit; however, the simultaneous degradation of both insulators intensified field distortion, with a maximum distortion rate of 25% at d = 20 mm. Increased detection distance led to a reduction in overall electric field strength and a lower distortion rate, indicating that the impact of electric field distortion diminishes with distance.

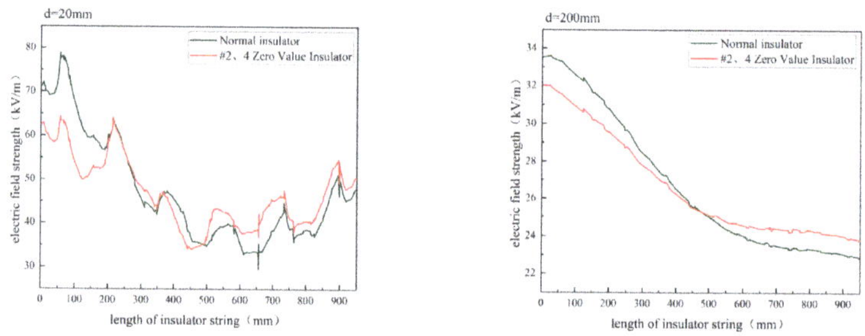

(**a**) (**b**)

Figure 6. Electric Field Distribution under Discontinuous Double Insulator Zero-Value Condition: (**a**) d = 20 mm; (**b**) d = 200 mm.

3.2.3. Electric Field Distribution Characteristics with Continuous Double Zero-Value Insulators

To investigate the influence of continuous dual zero-value insulators on the electric field distribution, Insulators #2 and #3 were set to zero-value, and electric field distribution measurements were taken at d = 20 mm and d = 200 mm, as depicted in Figure 7. Field strength was found to weaken below the zero-value area and strengthen above, with a smooth transition at the defect boundary. Continuous degradation caused more significant changes in the electric field than single-unit degradation, with a maximum distortion rate of 52.8% at d = 20 mm. The electric field intensity increased by approximately 20 kV/m near the third zero-value insulator compared to normal conditions.

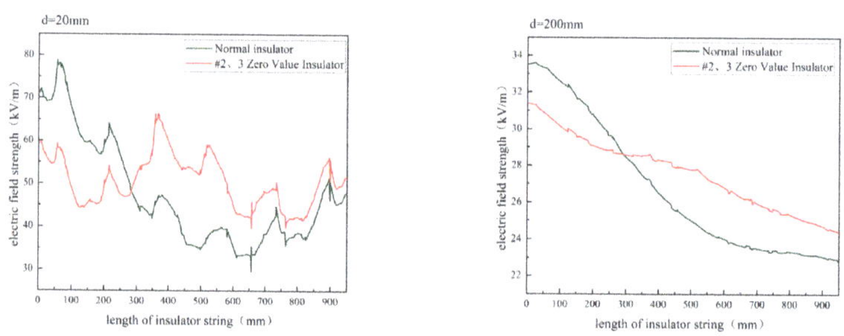

(**a**) (**b**)

Figure 7. Electric field distribution under continuous double insulator zero-value condition: (**a**) d = 20 mm; (**b**) d = 200 mm.

Comparative analysis of normal, single zero-value, discontinuous dual, and continuous dual degradation cases revealed that increasing detection distance reduces the impact of degraded insulators on the spatial electric field. As shown in Figure 8, degradation location significantly affects the spatial electric field, particularly at the high-voltage end and the tower connection, with less influence in the middle. The effect of continuous dual degradation on spatial electric field variation was found to be greater than that of discontinuous degradation and single-unit degradation.

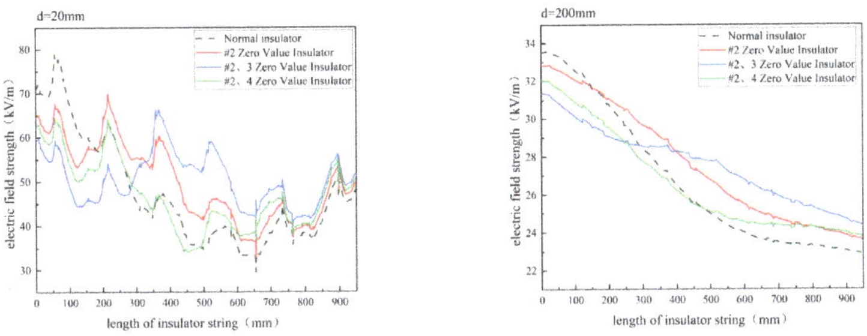

(a) (b)

Figure 8. Comparison of electric fields with left-side #1 and right-side #1 Zero-value insulators: (**a**) d= 20 mm; (**b**) d = 200 mm.

4. Intelligent Prediction Algorithm for Insulator String Electric Field Based on MLP Neural Network

Simulation data can only capture the electric field distribution characteristics at specific distances, making it challenging to comprehensively reveal variations across all distances. To facilitate on-site testing and data analysis, a multi-layer perceptron (MLP) neural network model was employed to construct a spatial electric field distribution dataset for insulator strings with zero-value defects. Based on this dataset, a corresponding database was established.

4.1. Multilayer Perceptron Neural Network Model

In this study, an MLP neural network combined with four different optimization algorithms was first employed to perform linear fitting of the target, establishing a three-layer neural network model to map the nonlinear relationship between electric field intensity and influencing factors. The MLP neural network was trained using the Levenberg–Marquardt algorithm, the Bayesian regularization algorithm, and the scaled conjugate gradient algorithm. Additionally, a genetic algorithm was embedded into the neural network to achieve global optimization of weights and thresholds. Finally, the established MLP neural network models based on the four optimization algorithms were applied to perform nonlinear fitting of the mapping relationship between electric field intensity and influencing factors. The convergence speed and fitting performance of each model were compared, and the Bayesian regularization-based MLP neural network was ultimately selected as the optimal model for predicting electric field intensity.

A three-layer MLP neural network model was employed to predict electric field intensity values based on two input parameters: the length of the insulator string (Insulator Length) and the distance of the measuring device from the insulator surface (Distance). The structure of the three-layer MLP model is illustrated in Figure 9, consisting of three primary components: an input layer, hidden layers, and an output layer. The input layer comprises two nodes corresponding to the two input variables: Insulator Length and Distance. The network features two hidden layers, each containing three neurons (also referred to as nodes). These neurons are interconnected with the input and subsequent layers through a set of weights (W). The primary function of the hidden layers is to perform complex nonlinear transformations on the input data, extracting higher-level feature representations from the raw data. The output layer contains a single node that represents and outputs the predicted electric field intensity value (Output) as the final result of the neural network's computation. This architecture highlights the neural network's robust capability to handle complex prediction tasks effectively.

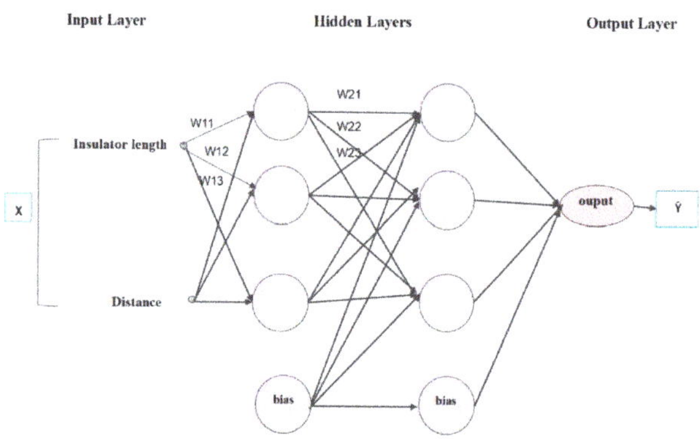

Figure 9. Three-MLP Neural Network Model.

Figure 10 illustrates the working principles of the Multi-Layer Perceptron (MLP) neural network, which can be systematically divided into four core stages: forward propagation, loss computation, backpropagation, and iterative training. During the forward propagation stage, input data are fed into the input layer and subsequently passed to the first hidden layer through predefined weights (W). Each node in a hidden layer receives weighted inputs from all nodes in the preceding layer and applies an activation function (such as sigmoid or ReLU) to determine whether the node is "activated." This activation mechanism propagates layer by layer until the output layer is reached, where the predicted electric field intensity value is generated. Next, the loss computation stage compares the predicted value with the actual electric field intensity using a specific loss function, such as mean squared error (MSE), to quantify the model's prediction accuracy. A smaller loss value indicates higher prediction precision.

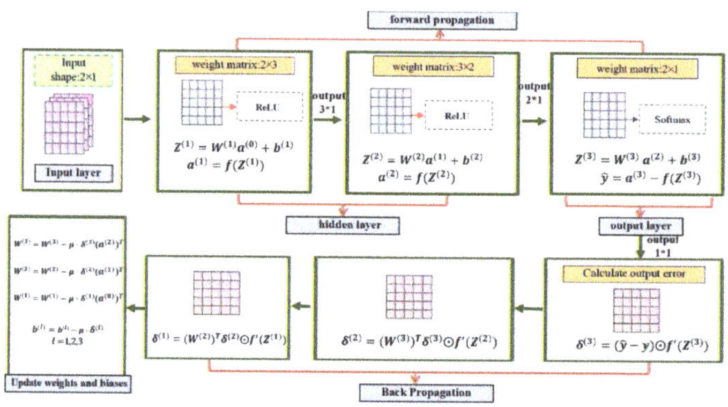

Figure 10. Data Processing Steps.

In the backpropagation stage, the gradient of the loss function is calculated, and optimization algorithms, such as gradient descent, are employed to iteratively adjust the network's weights (W) and biases (Bias). These adjustments aim to minimize the loss function value and improve the model's predictive performance. Finally, the aforementioned steps are repeated iteratively until the model converges, i.e., when the loss value decreases to a relatively small level. This process demonstrates the MLP neural network's capability to optimize predictive accuracy through iterative training.

The mathematical symbols in the figure are defined as follows:

Z: Weighted input vector.

W: Weight matrix connecting adjacent layers.
a: Activation output, where ReLU activation is applied to hidden layers and Softmax activation to the output layer.
b: Bias vector.
δ: Error term computed during backpropagation.
⊙: Hadamard product operator.

4.2. Analysis of Prediction Results

To evaluate the performance of the neural network model, commonly used regression metrics include:

Mean Squared Error (MSE): This metric quantifies the difference between the predicted and actual values. A smaller MSE indicates better predictive accuracy of the model.

$$\mathrm{MSE} = \frac{1}{\mathrm{K}}\sum_{\mathrm{i}=1}^{\mathrm{K}}(\mathrm{Ti} - \mathrm{Yi})^2 \quad (5)$$

where Ti represents the target value of the i-th sample, Yi is the predicted value, and K is the number of samples.

Coefficient of Determination (R^2): This metric reflects the model's explanatory power over the data. An R^2 value closer to 1 indicates a better model fit.

$$R^2 = \frac{\sum_{i=1}^{K}\left(Yi - \overline{T}\right)^2}{\sum_{i=1}^{K}\left(Ti - \overline{T}\right)^2} = 1 - \frac{\sum_{i=1}^{K}(Ti - \mathrm{Yi})^2}{\sum_{i=1}^{K}\left(Ti - \overline{T}\right)^2} \quad (6)$$

The model's predictions are evaluated using two metrics: MSE = 3.89457 and R^2 = 0.88. For the evaluation of prediction performance: an R^2 value greater than 0.8 indicates a good predictive capability of the model.

The predictive performance of the model can also be evaluated by generating relationship curves between insulator string length and electric field intensity. Taking a detection distance of 200 mm as an example, the comparison between the model's predicted values and simulation results is presented. Figure 11 illustrates the electric field intensity distribution of the seventh zero-value insulator, comparing simulation outcomes with model predictions. The trends of the simulation and prediction are largely consistent, demonstrating the model's reliability and validity, with errors not exceeding 10%.

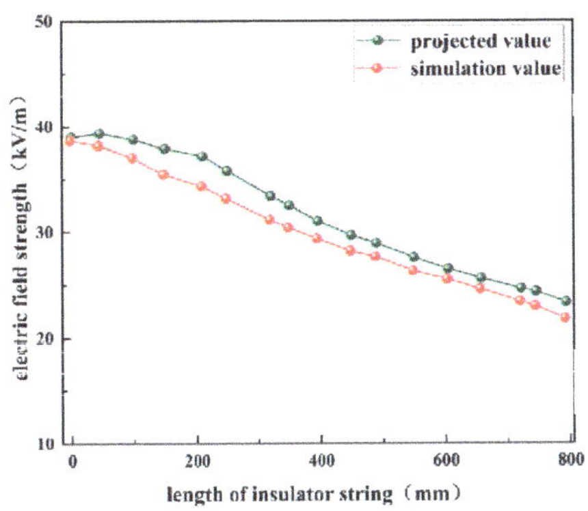

Figure 11. Comparison of model-predicted values and simulated values.

4.3. Experimental Validation of Prediction Outcomes

To verify the reliability of the proposed model, a degraded insulator detection test platform was constructed, as shown in Figure 12. A suspension string of nine XP-160 porcelain insulators was mounted on a cross-arm. The electric field detection device was placed inside a pod, which was secured using a pulley and a traction rope. The distance between the traction rope and the edge of the insulator string was maintained at approximately 200 mm. The detection device was uniformly pulled upward from the high-voltage end to the low-voltage end, measuring each insulator sequentially while recording the results. Degraded insulators were simulated by short-circuiting the steel caps and feet of porcelain insulators using conducting wires, with a degraded insulator placed as the third unit from the high-voltage end. The schematic diagram of the test wiring is provided in Figure 13. The voltage was supplied by the test power source, passed through a protective resistor, and applied to the high-voltage end of the insulator string via a capacitive voltage divider. The iron frame of the capacitive voltage divider and the low-voltage end of the insulator string were reliably grounded. Communication between the electric field detection device and the data-receiving terminal was established to ensure seamless data acquisition.

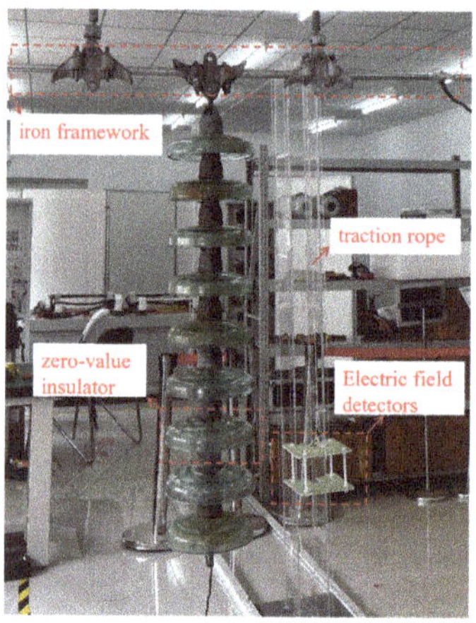

Figure 12. Laboratory Model of Insulator String.

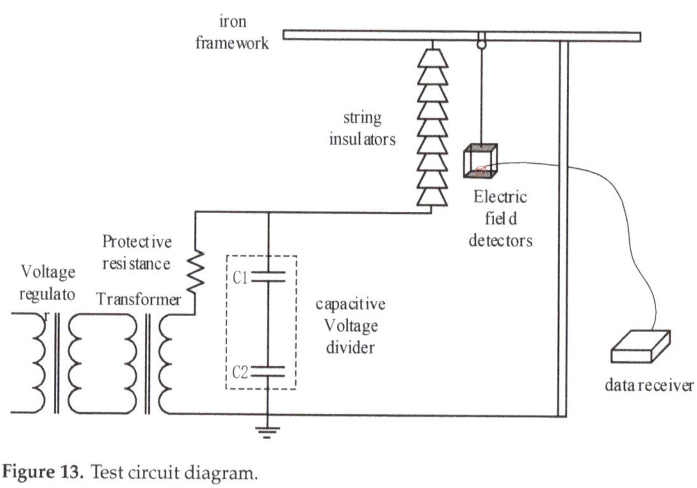

Figure 13. Test circuit diagram.

Due to the limitations of laboratory equipment, a voltage of 90 kV was applied in this experiment to simulate and evaluate the performance of degraded insulators on a 110 kV transmission line. The primary objective of the experiment was to validate the accuracy and effectiveness of the proposed model's predictions through empirical testing.

As shown in the data from Figure 14, the predicted electric field intensity at the first sampling point was 38.406 kV/m, while the corresponding experimental value was 37.9 kV/m, resulting in a calculated error rate of 1.31%. The maximum error occurred at the position where the insulator string length was 500 mm, with the predicted value being 25.356 kV/m compared to the experimental value of 27.357 kV/m, yielding an error rate of 7.32%. In contrast, the minimum error was observed near the second insulator, with an error rate of only 0.4%, Comparison between experimental and predicted values, as shown in Figure 15.

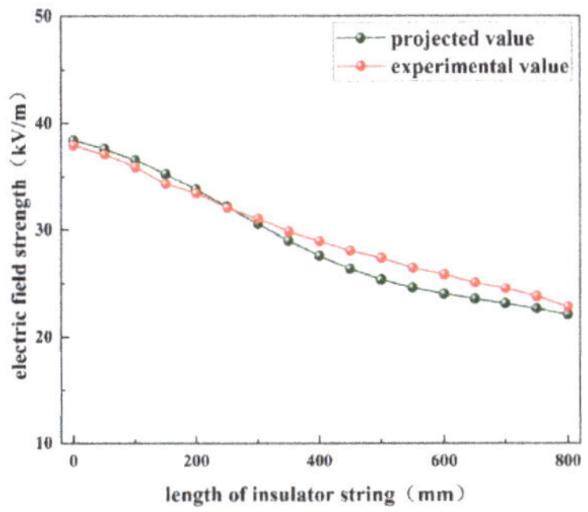

Figure 14. Comparison between experimental and predicted values.

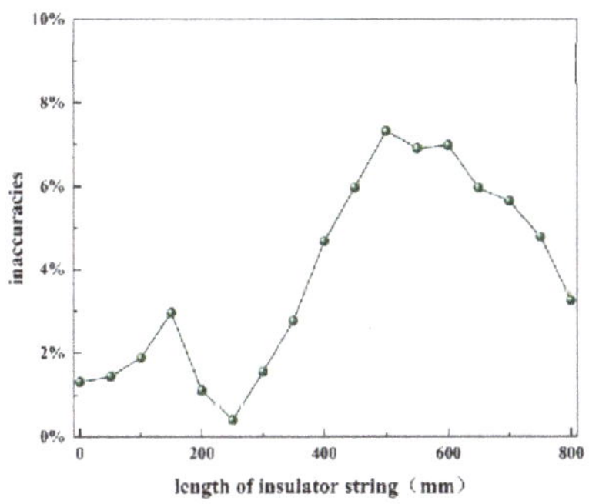

Figure 15. Error-values between experimental and predicted values.

A comparison of the model's predictions with experimental results reveals a high degree of consistency in overall trends, with all errors remaining within 10%. This finding partially validates the reliability of the model. However, it is noteworthy that significant deviations in electric field intensity were observed at different distances, likely due to discrepancies in insulator dimensions and experimental environmental conditions.

Further analysis reveals unavoidable differences between the experimental environment and the tower line and hardware structure set in the simulation model. As a result, there is a noticeable amplitude deviation between the actual measured and predicted electric field intensities at the same detection distance. These deviations highlight the need to thoroughly account for environmental factors affecting electric field distribution before applying the model to real-world engineering scenarios, necessitating appropriate calibration and optimization of the model.

5. Conclusions

This study employs electrostatic field simulation based on COMSOL to investigate the effects of detection distance, position, and quantity of zero-value insulators on electric field distribution. The spatial electric field distribution patterns of insulators were examined under various conditions and distances, leading to the following conclusions:

(1) A detection distance of 20 mm allows for a significant reflection of the electric field strength variations near the porcelain insulator string. However, when the detection distance exceeds 300 mm, the response to zero-value insulators is diminished. Therefore, in practical engineering applications, it is recommended that the detection distance be controlled within the range of 20 mm to 300 mm.

(2) In the three-dimensional electric field zero-value condition, the impact of degraded insulators diminishes with increasing measurement distance. The effect is more pronounced at the high-voltage end and the tower connection end, with minimal influence at the center. Additionally, continuous degradation of two insulators has a greater impact on the spatial electric field variation rate than intermittent or single insulator degradation.

(3) To construct a spatial electric field distribution database for insulator strings with zero-value defects, a Multi-Layer Perceptron (MLP) neural network model was employed in this study. Its predictive accuracy was thoroughly validated through rigorous experiments. This database provides robust data support for insulator degradation detection, thereby enhancing the precision and reliability of insulator condition monitoring and analysis.

Author Contributions: Conceptualization, P.Y.; formal analysis, P.Y., L.Z., J.L., H.L. (Hui Liu) and H.L. (Hao Luo); Investigation, P.Y., L.Z., J.L., H.L. (Hui Liu), T.L. and H.L. (Hao Luo); writing—original draft, T.L., H.L. (Hui Liu) and H.L. (Hao Luo); writing—review and editing, P.Y., H.L. (Hui Liu) and H.L. (Hao Luo). All authors have read and agreed to the published version of the manuscript.

Funding: This research was funded by the State Grid Corporation of China (SGCC) Headquarters Science and Technology Projects, grant number 5500-202316523A-3-2-ZN.

Data Availability Statement: The data presented in this study are available on request from the corresponding author.

Conflicts of Interest: Authors Lei Zheng, Pengxiang Yin, Jian Li, Tao Li and Hao Luo were employed by the companies NARI Group Corporation Ltd. and Wuhan NARI Limited Liability Company. The remaining authors declare that the research was conducted in the absence of any commercial or financial relationships that could be construed as a potential conflict of interest.

Abbreviations

The following abbreviations are used in this manuscript:

MLP Multilayer Perceptron

References

1. Miao, C.; Wang, W.; Wang, R.; Chen, G.; Li, J.; Tang, W. Finite Element Analysis of Fracture and Size Optimization of Steel Pin Ball of Cap and Pin Type Suspension Insulator. *Insul. Surge Arresters* **2022**, *1*, 165–170.
2. Ying, X.; Zhang, M.; Wang, S.; Qin, L. Design of a non-contact fault detection system for porcelain insulators based on laser excitation. *Electron. Des. Eng.* **2021**, *29*, 10–14.
3. Li, J.; Lü, Y. Status Quo and Prospect of Research on Contamination Characteristics of Insulators. *Guangdong Electr. Power* **2018**, *31*, 18–23.

4. Zeng, L.; Zhang, Y.; Zeng, X.; Li, T.; Deng, Z.; Wang, P.; Wan, H.; Xu, B.; Liu, Y.; Dong, C.; et al. Aging State Evaluation Methods for Silicone Rubber Sheds of Composite Insulators. *Insul. Surge Arresters* **2022**, *2*, 139–145+152.
5. Li, Y.; Zhou, L.; Li, T.; Shao, X.; Chu, J.; Gao, L. Fault Analysis for Lightning Strike-caused String Fracture of 500 kV Line Porcelain Insulators. *Zhejiang Electr. Power* **2020**, *39*, 17–21.
6. Jiang, Y.; Han, J.; Ding, J.; Fu, H.; Wang, Y.; Cao, W. The identification and Diagnosis of Self-Blast Defects of Glass Insulators Based on Multi-Feature Fusion. *Electr. Power* **2017**, *50*, 52–58+64.
7. Zhang, Q.; Li, C.; Sun, Y.; Yang, C. Accurate Identification of Insulator Based on Deep Learning. *Insul. Surge Arresters* **2022**, *1*, 143–150.
8. Zhou, Q. Sequence of Live Testing on Suspension Insulators Using Spark Gap Methodology. *Electr. Saf. Technol.* **2002**, *3*, 25–26.
9. Chi, D.; Zeng, Q.; Sun, L. Detecting Insulator by UV Imaging. *High Valtage Eng.* **2006**, *2*, 115–116.
10. Cheng, Y.; Xia, L.; Li, Z.; Cheng, D.; Yan, B.; Li, S. Detection of Zero-Value Ceramic Insulators Based on Infrared Imaging Methodology. *Insul. Mater.* **2019**, *52*, 74–79.
11. Wang, S.; Jiang, T.; Li, W.; Lü, F.; Liu, P. Study on Infrared and UV Imaging Characteristics Test and Coupling Field Simulation of Zero-value Insulator Strings. *High Volt. Appar.* **2021**, *57*, 1–9.
12. Ramesh, M.; Cui, L.; Gorur, R. Impact of superficial and internal defects on electric field of composite insulators. *Int. J. Electr. Power Energy Syst.* **2019**, *106*, 327–334. [CrossRef]
13. Cheng, Y.; Li, C.; Ma, X. Study on Online Detection of Faulty Composite Insulators by Electric Field Method. *High Volt. Eng.* **2002**, *S1*, 8–63.
14. Xu, Y.; Lu, J.; Zhu, X.; Zhang, C.; Zhang, D.; Liu, X.; Huang, X. A Noncontaet Zero-value Insulator Detection Method and Device. *Guangdong Electr. Power* **2021**, *34*, 130–136.
15. Zhang, D.; Chang, Z.; Wan, W.; Zhang, Z.; Chen, J. Zero-value insulator detection technology based on local electric field. *Eletr. Power Eng. Tehnol.* **2024**, *43*, 193–201.
16. Liu, P.; Wu, Z.; Zhu, S.; Xu, J.; Liu, Q.; Peng, Z. Simulation on Eletric Field Distribution of 1 100kY AC Tri-Post Insulator Influenced by Defects. *Trans. China Electrotech. Soc.* **2022**, *37*, 469–478.

energies

MDPI

Review

The Progress and Prospect of Gap Breakdown Characteristics and Discharge Mechanisms of Overhead Transmission Lines Under Vegetation Fire Conditions

Haohua Hu [1], Peng Li [2,*] and Daochun Huang [2,3]

[1] College of Electrical Engineering and New Energy, China Three Gorges University, Yichang 443002, China; 202208540021007@ctgu.edu.cn
[2] Transmission Line Engineering Technology Research Center of Hubei, Yichang 443002, China; huangdc99@163.com
[3] School of Electrical Engineering and Automation, Wuhan University, Wuhan 430000, China
* Correspondence: lipeng_ctgu@163.com

Abstract: Wildfires frequently occur, posing a significant threat to the operational stability of transmission lines across mountainous forest areas. Therefore, this paper reviews numerous studies conducted by domestic and international scholars on the gap breakdown tests and discharge mechanisms of transmission lines under simulated wildfire conditions. It analyses and summarizes the physical parameter measurement methods commonly used in current experiments. Combining the results of existing experiments, this study analyzes the discharge mechanisms, including the research progress made in numerical simulations. The conclusion is that existing tests are limited in their measurement methods of the physical quantities related to breakdown characteristics, and it is not easy to strictly control experimental variables when considering complex factors. Numerical simulations mainly focus on multi-physical field simulations, which consider the characteristics of vegetation fires in short gaps. The synergistic mechanism of environmental factors on gap breakdown characteristics remains unclear. This paper points out the breakdown characteristics and discharge mechanisms derived from existing experiments and numerical simulations under various influencing factors, highlighting their applicability and limitations, which differ from complex actual transmission lines in the environment. Then, we look forward to the future development of simulation test platforms that could better reflect the actual transmission line corridor environment, incorporating multi-parameter measurement and in-depth numerical simulation works that consider climate and terrain factors.

Keywords: wildfire; transmission line; HVDC; gap breakdown; parameter measurement methods; breakdown characteristics; discharge mechanism; numerical simulation; predictive model

1. Introduction

The distribution pattern of energy resources and electricity load determines the development demand for long-distance, high-capacity transmission in China, making it inevitable that transmission line corridors will need to pass through areas prone to wildfires. Wildfires, due to their inherent combustion characteristics, conductivity, ash, and particles, along with the complex interaction of environmental factors such as variable wind speeds and intricate terrain in line corridors, can significantly reduce the insulation strength of transmission line gaps, leading to line tripping and low reclosure success rates, which severely threaten the reliability of transmission line operations.

According to the statistics from the China Southern Power Grid, forest fires in regions such as Yunnan, Guangxi, Guangdong, Hainan, and Guizhou within its jurisdiction account for over 80% of the national total yearly [1]. The distribution statistics of line tripping caused by wildfires in various jurisdictions from 2015 to 2020 are as follows: Yunnan 88 times, Guizhou 68 times, Guangxi 59 times, Guangdong 73 times, Hainan 54 times, and the UHV company 35 times (accounting for 9.3%). The amount of line tripping caused by wildfires in each provincial grid was over 50 times, with Yunnan having the highest tripping rates, accounting for 23.3% of the total number in the Southern Power Grid jurisdiction [2]. Wildfire remains one of the primary causes threatening the safe and stable operation of transmission lines.

Analyzing the breakdown characteristics and discharge mechanisms of air gaps under wildfire conditions is essential for transmission line risk assessments and early warnings. As shown in Table 1, experimental simulations using conductor–plate configurations and woodpile fires enable the preliminary replication of wildfire scenarios, allowing for the measurement of breakdown voltage and leakage current under varying conditions. However, key microscopic parameters such as charged particle distribution and electric field strength, which are difficult to observe experimentally, can be effectively analyzed using numerical simulation methods.

Table 1. Comparison of existing research methods.

Methods	Advantages	Disadvantages
Experimental simulation	Real-world environment conditions.	Susceptible to extreme weather, especially in outdoor long-gap tests.
	Woodpile combustion with good repeatability.	Difficult to capture microphysical phenomena during discharge.
	Direct observation of gap discharge phenomena and data recording.	Challenging to isolate and control single variables accurately.
Numerical simulation	Transformed into solvable multi-physics coupling problems.	Simplified model assumptions with deviation.
	Capable of simulating microscopic parameters.	Less directly applicable simulation conclusions.
	Reveal underlying discharge mechanisms.	A lack of sufficient experimental reference data for accuracy and reliability.
Contributions		
Experimental and numerical methods are mutually supportive. Experiments provide key discharge data for model development and validation, while simulations offer mechanistic insights into microparameters beyond experimental research.		

This paper reviews the current research on gap breakdown characteristics and discharge mechanisms under simulated wildfire conditions, both domestically and internationally, and emphasizes their applicability and limitations relative to real-world transmission line environments. Then, it looks ahead to future research on constructing realistic experimental platforms that simulate composite environmental factors. The aim is to achieve comprehensive parameter measurements and multi-physics coupling gap discharge models.

2. Materials and Methods

2.1. Research on Breakdown Characteristics

2.1.1. Analysis of the Factors Affecting Power Line Tripping Caused by Wildfires

Figure 1 shows the primary research approach for gap breakdown characteristics and discharge mechanisms under vegetation fire conditions. Due to the high cost, long duration, and poor repeatability of conducting large-scale wildfire breakdown characteristic tests on actual transmission lines, current research mainly focuses on simulating wildfire gap conditions.

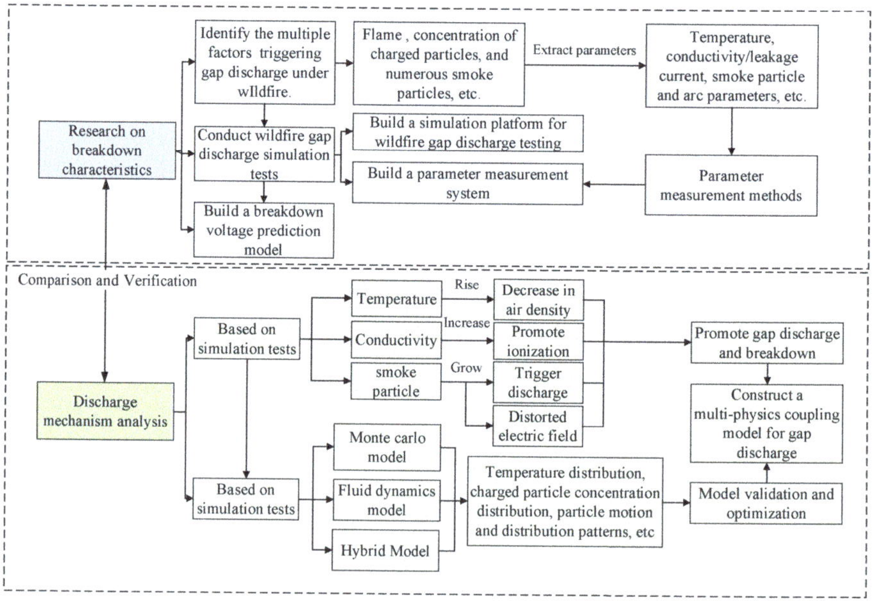

Figure 1. Mind map of gap breakdown characteristics and discharge mechanisms of overhead transmission lines under vegetation fire conditions.

The gap breakdown characteristics of transmission lines under wildfire conditions involve numerous and complex influencing factors, with each factor being highly coupled and interacting synergistically in the discharge process of the flame gap [3]. Based on the research outcomes of domestic and international academic teams, key influencing factors have been summarized, including temperature, conductivity, and smoke particles. Some researchers have also conducted numerous experiments focusing on factors such as vegetation type, wind speed, slope, and altitude [4–6].

2.1.2. Parameter Measurement Methods

In response to the above-influencing factors, the physical parameters involved in wildfire simulation experiments include temperature, conductivity, smoke particle parameters, and arc parameters.

1. Temperature measurement

Thermocouples are commonly used for temperature measurements in wildfire simulation experiments. Lu Wei and others conducted combustion characteristic experiments on common vegetation types in transmission line environments, using K-type, armored thermocouple trees to collect temperature data at different flame heights. The maximum temperatures recorded for the combustion of $21 \times 21 \times 10$ cm^3 piles of cedar, pine, and eucalyptus were 747 °C, 719 °C, and 687 °C, respectively [7]. The thermocouple temperature measurement is a common contact-based method. In recent years, non-contact temperature measurement techniques have developed rapidly. These methods do not physically contact the measured temperature field, avoiding interference. They offer advantages such as non-invasiveness, fast response times, high-resolution, and real-time measurements, making them promising methods for application in high-voltage breakdown tests. Table 2 compares the advantages and disadvantages of thermocouples with several non-contact temperature measurement methods.

2. Conductivity Measurement

The measurement of conductivity also involves the collection of leakage current. Measurement methods include microwave measurement and local average conductivity, among others.

Lu Wei et al. conducted conductivity characteristic experiments under high-risk vegetation fire conditions. They collected leakage current using a sampling resistor module, data acquisition card, and PC-based system. They found that the local conductivity of the fir pile flames was approximately 30 μS/cm, with conductivity decreasing as the flame height increased [7]. Due to its low-temperature weak plasma, the local conductivity of flames can also be measured using laser interferometry, which offers high spatiotemporal resolution. This method allows for the plasma refractive index, electron density, expansion evolution profile, and expansion velocity, among other state parameters, to be determined. A Mach–Zehnder interferometer splits a laser beam into two coherent beams using a beam splitter. One beam serves as the reference light, and the other interferes with the probe light that passes through the plasma to be measured, with the resulting image captured by a CCD camera. The phase changes in the plasma are extracted from the interference pattern obtained, and the spatial refractive index of the plasma is obtained using Abel transform. Local conductivity and other microscopic parameters can be diagnosed based on the intrinsic relationship between the refractive index and plasma [8].

3. Smoke Particle Parameter Measurement

Smoke particle parameters include smoke concentration, particle size, and the particle charge-to-mass ratio.

Particle concentration can be measured using smoke sensors or lidar monitoring systems. Yang Nongchao et al. studied the effect of vegetation combustion smoke on the gap breakdown strength under voltages of 10–35 kV. They installed a smoke sensor 1 m above the conductor to analyze the effects of smoke alone [9].

Smoke, as an aerosol, undergoes Mie scattering and attenuation when laser light strikes its surface, generating optical signals that reflect the physical state of smoke. These signals are then converted to electrical signals to calculate smoke concentration [10]. Liu Xinyang et al. designed an experimental platform for measuring smoke concentrations composed of a pulsed laser emission system, smoke chamber, and pulsed laser reception system. They tested the scattering characteristics of four typical types of smoke, including beechwood and incense, and used the Winner318 laser particle size analyzer to measure the smoke particle size distribution [11].

The particle charge-to-mass ratio refers to the ratio of the charge carried by the particle to its mass. Wang Ziming et al. built an experimental platform with which to measure the particle charge-to-mass ratio. The equipment for measuring the particle charge includes a Faraday cup and a high-precision charge meter, while the particle mass is measured using a high-precision electronic balance. The measured charge-to-mass ratio for 20-mesh particles showed the slightest relative error of 16.96% compared to the theoretical value, demonstrating a higher measurement accuracy for larger particles [12].

Table 2. Comparison of various temperature measurement methods.

Temperature Measurement		Principle	Structure	Advantages	Disadvantages
Contact	Thermocouple, thermistor, and expansion thermometer [13]	Using temperature measurement elements in direct contact with the object to be measured, reaching thermal equilibrium.	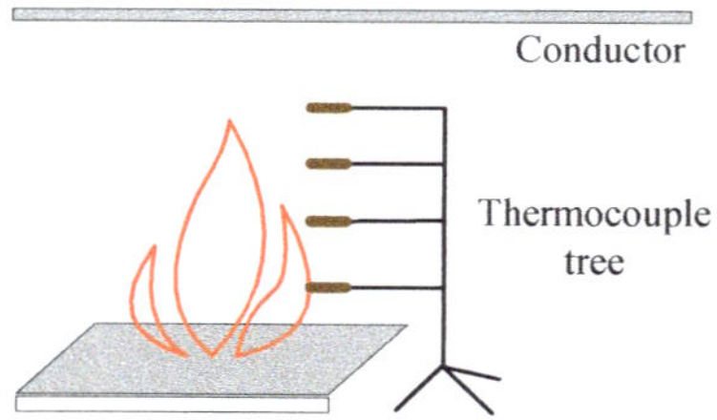	Simple structure, easy to use, highly reliability, and can directly obtain the variation curve of flame temperature with time and height.	Interference with the measured flow field achieved, slow response speed, and not suitable for transient temperature measurement.
Non-contact	Acoustic method [14]	The propagation speed of sound waves in a medium is functionally related to its temperature and type.	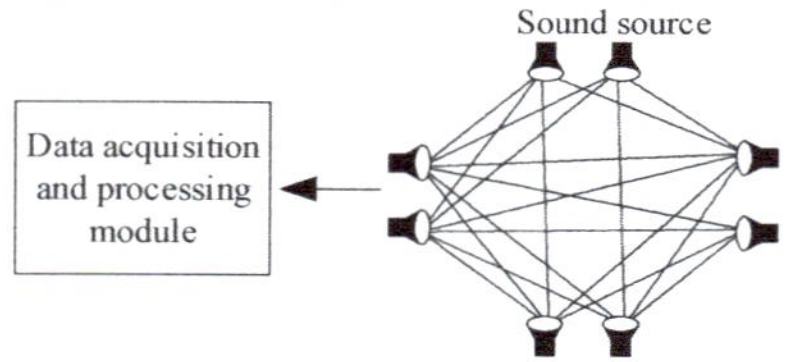	The response speed has been improved.	Easily affected by dust, particulate matter, and airflow.
	Infrared radiation [14]	Identifying and measuring the thermal radiation spectrum of an object.		Fast response speed, high resolution, and capable of providing 2D and 3D flame temperature distributions.	Unable to provide dynamic monitoring of the measured object and work in rainy conditions.
	Holographic interference [15,16]	Using light interference passing through the measured medium, the temperature distribution is obtained by processing recorded light information.	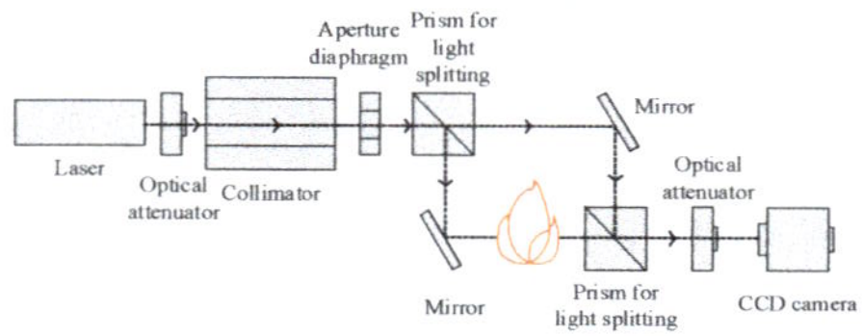	Suitable for quantitative and qualitative measurements of dynamic objects in real-time measurement.	Includes optical instruments with complex experimental setups for measuring 3D turbulent flames.

4. Arc Parameter Measurement

Arc parameters include electrical, morphological, and microscopic parameters. Electrical parameters include the discharge initiation phase flow path and the electric field distribution at the arc head. Morphological parameters include the arc radius, development speed, length, and offset angle. Microscopic parameters include the density distribution of charged particles within the flame and their dynamic evolution.

Currently, the focus on arc parameters is mainly concentrated on their macroscopic morphological characteristics, primarily recorded by cameras. Using ICCD cameras with nanosecond shutter speeds, the morphology of the flowing arc can be captured, and the development rate of the arc can be estimated [17]. Many high-speed cameras available on the market can meet the experimental requirements for studying the arc path in vegetation fire gap breakdowns. In addition to using ICCD cameras, relevant scholars have also measured the development speed of the flow arc using photomultiplier tubes (PMTs) and scanning cameras. Grangel calculated the average development speed of the dry air gap flow arc at atmospheric pressure by analyzing the signal time differences collected at different locations using a PMT. The resulting speed is typically on the order of 0.01 mm/ns [18]. Sigmond used scanning camera technology to study the flow discharge speed of a needle-plate electrode gap of 0.5 cm and 1 cm, finding that the flow speed gradually slowed as it moved from the needle electrode toward the plate electrode, eventually reducing to 1.4 mm/ns when it contacted the plate electrode [19]. Existing research mainly focuses on measuring the flow arc development speed in short gaps in pure air and has not profoundly discussed the impact of firelight in the gap on the detection efficiency of photodetectors. Therefore, the direct application of ICCD cameras to arc observation in simulated wildfire environments has limitations.

Flow discharge's electric field strength measurement tends to use optical field sensors based on the electro-optic effect. Hidaka from the University of Tokyo developed a sensor utilizing the Pockels effect to measure the electric field of positive flow discharge in a 10 cm needle-plate gap [20,21]. The research team from Tsinghua University developed a photodetector sensor based on the Mach–Zehnder interference principle to measure the spatial electric field of flow discharge. However, when the flow channel extends near the sensor, oscillations in the detected electric field occur, indicating that optical field sensors still have limitations in measuring the internal electric field of the flow discharge [22,23].

Currently, the development direction of parameter measurement technology is mainly used to meet the demands of high detection accuracy, multi-environment adaptability, and cost-effectiveness. Under simulated wildfire conditions, environmental factors include high flame temperature, strong spatial electric fields, and abundant ash smoke. Regarding the measurement of macro parameters such as temperature, conductivity, and smoke, traditional close-contact parameter measurement systems require high standards for equipment's high-temperature resistance, insulation performance, and dust prevention measures. On the other hand, constructing a remote non-contact measurement system involves considering the overall layout of the experimental platform and space utilization, which increases the design complexity and cost. There are limited methods for measuring microscopic parameters such as charged particle concentration, arc development speed, and background electric field strength, and there is a lack of cases applied to simulated experiments.

3. Results

3.1. Breakdown Characteristic Experiments

Foreign studies on the breakdown characteristics of wire-to-ground gaps under simulated wildfire conditions began early, as shown in Table 3. The applied voltage types include both AC and DC conditions, and the gap distances range from centimeter-level short gaps to several meters in scale in the simulated full-scale tests. However, early simulated breakdown tests considered relatively few influencing factors, had limited testing methods, and exhibited poor reproducibility of the test fire sources. As a result, the average breakdown voltage values obtained by different scholars have varied significantly.

Table 3. Summary of international research status on the line gap breakdown test under simulated wildfire conditions.

Scholars	Factors	Gap Distance/m	Fire Source	Main Conclusions
Chun K [24]	Heat release rate and effective combustion heat	Within 1	Cypress, fir, and two types of plastic resins: PP and PE	Gap dielectric strength can be predicted based on the ignition characteristics of vegetation and building materials
Z. Ntshangase [25]	Flame temperature	0.5	Sugarcane	Under negative polarity voltage, the breakdown voltage under flame conditions decreases by 50%, whereas the number is 55% under positive polarity voltage.
A. Robledo-Martinez [26]	Flame temperature, electrical conductivity, and smoke particles	0.85~2.0	Horticultural waste, sugarcane bagasse, branches and leaves, and butane gas	The main factors for gap insulation performance are high temperature, the ionization of multiphase component mixtures, and smoke particles.
Fonseca [27]	Type of vegetation	1	Grass and sugarcane	The insulation requirements for AC transmission lines were established through the test.
Moreno [28]	Non-fire and high-temperature flames	3	Sugarcane	The insulation strength under gap flame conditions is significantly weakened compared to a purely high-temperature, non-flame environment.
Lanoie [29]	Gap distance	6	Eucalyptus	The gap breakdown voltage gradients with and without considering the eucalyptus tree's height were 32.8 kV/m and 58.4 kV/m, respectively.

The teams from Wuhan University and the University of Science and Technology of China have conducted extensive research on the breakdown characteristics of gaps under simulated wildfire conditions, considering factors such as flame temperature, high conductivity, numerous smoke particles, wind speed, slopes, and other terrain influences. Table 4 summarizes the simulated wildfire gap breakdown tests conducted by different research teams, addressing gaps of various scales and influencing factors. Significant progress has been made in monitoring changes in breakdown voltage and leakage current, arc breakdown paths, and altitude corrections. These advancements have provided a wealth of reference data for further understanding the gap discharge mechanism.

Table 4. Summary of domestic research status of line gap breakdown tests under simulated wildfire conditions.

Scholars	Factors	Gad Distance/m	Fire Source	Main Conclusions
Huang, D. [30–33]	Vegetation type and combustion intensity	0.45	Cedar (trunk and branches), straw, and reed grass	The peak leakage current in the positive polarity of the cedar branch flame gap (27 mA) is much higher than that under a negative polarity voltage (0.6 mA).
	Ash particles	0.40, 0.45	Reed grass, straw, and cedar branches	The average breakdown voltage gradient of the conductor-to-plate gap (1.4 kV/cm) decreases by 71% compared to that under pure air conditions (4.5 kV/cm).
	Flame-bridging ratio	1.2, 1.7, 2.2, 2.7	Cedarwood stack	The flame region is divided into continuous, discontinuous, and smoke zones based on the flame morphology and its impact on discharge characteristics.
	Vegetation moisture content	3	Cedarwood stack	The linear fitting formula was obtained between the average gap breakdown voltage gradient and vegetation stack humidity, with a high correlation coefficient of 0.9998.

Table 4. *Cont.*

Scholars	Factors	Gad Distance/m	Fire Source	Main Conclusions
Li, P. [4,34]	Wind speed	0.40~0.80	Cedar (branches and trunk), straw, and reed grass	When vegetation combustion is sufficient, wind speed significantly affects the flame-bridging gap ratio, with the breakdown voltage increasing as wind speed rises.
	Slope			The breakdown voltage increases by 32% as the slope rises from 0° to 19°, and the breakdown location shifts toward the upper-middle part.
Long, M. [5]	Ash particles and vegetation type	0.40, 0.45, 0.50, 0.55	Reed grass, straw, cedar leaves, and branches	The effect of particles on discharge development is positively correlated with particle size.
Yang, K. [35]	Temperature, conductivity, and ash particles	1	N-heptane and wood stack blocks	The breakdown field strength decreases to 9.5% at a gap of 10–50 cm and 3.7% at 30–70 cm in pure air, respectively.
Zhou, E. [6,36]	Altitude	1.2~1.7	Cedarwood stack	A method for the altitude correction of air gap breakdown voltage under vegetation fire conditions is proposed.
	Vegetation types, combustion characteristics, etc.	1.7, 3, 4	Metasequoia, Yunnan pine, fast-growing eucalyptus, shrubs, and reed grass	The average voltage gradient ratios of Yunnan pine, fast-growing eucalyptus, shrubs, and reed grass relative to Metasequoia were 1.13, 0.96, 0.93, and 0.88, and used as vegetation characteristic parameters.
Chen, X. [37]	Flame height	1	Cedarwood stack	The breakdown voltage is not strictly negatively correlated with flame height.
Wang, T. [38]	Smoke-bridging ratio	4	Cedarwood stack	The breakdown voltage under full smoke conditions is 582.25 kV, which is 45.59% of that under pure air conditions.

Based on the above research, it is evident that the primary method for studying the breakdown characteristics of transmission line gaps under wildfire conditions remains the simulation of wire-to-ground gaps and the generation of vegetation fires using vegetation stacks. However, the vegetation stacks arranged in a regular, equal-volume pattern neglect the impact of the actual growth shape of tree canopies on flame characteristics and discharge properties, such as the conical shape of mature fir trees and the pagoda-like shape of pine trees. Different canopy shapes and vegetation densities shield the electric field beneath the lines, affecting the surrounding electric field distribution [39,40]. At the same time, the current experimental platforms cannot meet the real-time multi-parameter observation requirements during the gap discharge process in actual wildfires. This undoubtedly increases the overall test volume and makes acquiring and analyzing more parameters concurrently under the same operating conditions challenging. There is still a lack of effective variable control methods for smoking concentration, particle size, and the gap-bridging ratio. The mechanisms of environmental wind speed, complex terrain slopes, and different altitudes that affect gap breakdown characteristics still require further study. Large-scale simulated tests for transmission line gap distances under higher voltage levels, such as UHVDC and UHVAC, are relatively rare. This makes providing adequate data support and guidance for insulation failure risk assessment and dynamic defense work for UHV transmission lines under wildfire conditions difficult.

3.2. *Breakdown Voltage Prediction Model*

As shown in Table 5, scholars have proposed empirical or semi-empirical formulae for the breakdown voltage of gaps under wildfire conditions based on various simulated experimental data and numerical simulation results both domestically and internationally [41].

Table 5. Summary of research status of breakdown voltage prediction models.

Scholars	Prediction Model	Key Physical Quantities	Applications and Conclusions
Imtiaj Khan, Mona Ghassemi [42]	$\hat{U}_{50} = U_{50} \times C_T \times C_S$ $U_{50} = 912.75log(X) + 640.8$	Temperature correction factor C_T	A steady-state probability model of line interruption is established considering breakdown voltage, line tripping, and aging failure.
		Particle correction factor C_S	
Zhou, E. [36]	$y_{100} = 2.19x_1 - 0.02x_2 - 1.89x_3 - 4.54x_4 + 164.88$ $y_{75} = -18.34x_1 - 0.016x_2 - 10.15x_3 - 6.15x_4 + 240.03$	y_{100} and y_{75}: Average breakdown voltage gradient at 100% and 75% flame-bridging ratios.	The average voltage gradient prediction models under flame-bridging ratios of 100% and 75% are obtained, with a coefficient of determination of 1 and good data fitting.
		x_1: Flame height; x_2: Maximum flame temperature; x_3: Ash content by mass; x_4: Calorific value of combustible material.	
Li, P. [34]	$U_{ap} = \frac{E_x x}{C_{ng}} + \frac{E_H H}{C_a C_b C_g}$	E_x and E_H: Average breakdown voltage gradients in the non-flame and flame zones (kV/m).	The breakdown voltages of different flame gaps are predicted, with an error between the predicted and actual values of less than 7.8%, validating the feasibility of the prediction model.
		x and H: Gap lengths in the non-flame and flame zones (m).	
Shao, Y. [34,43,44]	$U = \frac{h}{H} K_t K_{p1} U_0 + (1 - \frac{h}{H}) K_{p2} U_0$ $K_t = K_d K_h$	U: Line frequency breakdown voltage under wildfire conditions; U_0: Line frequency breakdown voltage under standard atmospheric conditions.	The breakdown characteristics of a typical 1000 kV ultra-high voltage double-circuit line are analyzed, and the main factors of gap insulation strength are comprehensively considered.
		K_t, K_d, and K_h: Atmospheric, air density, and humidity correction factors.	
		K_{p1} and K_{p2} are the particle correction factors for the flame zone and non-flame zone.	
Zhou, Z. [3]	$U_f = \int_{z_{al}}^{D_1 + z_d} E_0 \cdot \frac{T_a}{T_f} dz$	T_a: Ambient temperature (K); A: Flame height conversion coefficient; z_d: Vegetation height (m); T_f: Flame zone temperature (K)	Breakdown voltage prediction models are proposed for each different flame zone by height, considering mathematical relationships with vegetation, wind speed, and terrain parameters.
		E_0: Breakdown field strength in standard atmospheric conditions; D: Gap distance (m); D_1: Height of the flame zone (m).	

Model validation and error estimation are critical steps in constructing breakdown voltage prediction models, serving as key references for assessing model accuracy and applicability. Common validation methods include comparing experimental data and analyzing real transmission line case studies. Models are refined based on prediction outcomes, and targeted follow-up studies are conducted to improve their completeness and accuracy. The prediction model developed by Li Peng et al. achieved errors within 7.8% between predicted and measured values under different vegetation types and heights. This model was preliminarily applied to 220 kV and 500 kV systems with gap lengths of 6.5 m and 11 m, respectively, but it could not predict breakdown conditions dominated by herbaceous vegetation ash [34]. Shao Youguo et al. proposed a prediction model incorporating flame zoning and particulate corrections and applied it to 1000 kV UHV and 500 kV EHV cases. They found that UHV systems were more susceptible to inter-phase breakdown under the same smoke concentration compared to EHV systems [44].

These theoretically derived breakdown voltage models often fail to fully account for the correction effects of various influencing factors or validate the model while neglecting specific variables. Consequently, their applicability to predicting wildfire-induced tripping probabilities on real transmission lines remains limited.

Many studies have considered breakdown voltage prediction under high temperature, conductivity, smoke particles, and correction factors for different vegetation types. The breakdown voltage prediction formulae proposed within the context of their respective experiments offer some reference value for specific conditions. However, the transmission line corridor in an actual wildfire environment, in addition to the above factors, also includes real-time varying micro-meteorological data, such as environmental wind speed, precipitation, and humidity, as well as the complex terrain of specific regions, such as slope, slope orientation, and unique topographical features, all of which must be considered for their impact on the gap breakdown voltage. At the same time, the many influencing factors involved in the transmission line tripping process triggered by wildfires are interrelated. While prediction models built under independent conditions of each factor show good consistency in experimental studies, they are challenging to apply to actual line-tripping warnings under wildfire conditions. Therefore, constructing breakdown voltage prediction models under the influence of various factors requires a balance between comprehensiveness and coupling.

3.3. *Discharge Mechanism Analysis*

3.3.1. Mechanism Analysis Based on Simulated Experiments

Based on simulated experimental data and the derivation of breakdown prediction models, combined with the theoretical analysis of extended air gap discharge under uneven electric fields, scholars both domestically and internationally have proposed three basic theoretical models that consider the decrease in air density, the increase in flame conductivity, and particle-triggered discharge, as shown in Table 6, to reveal the discharge mechanism of gaps in vegetation fires.

Table 6. Mechanism of vegetation fire gap discharge.

Factors	Key Mechanisms	Modes of Action
Temperature [45]	Decrease in air density	Positive correlation between the gas's relative density and the gap's breakdown voltage
Conductivity [35,46,47]	Generating charged particles.	Collisional ionization and photoionization
		Thermal ionization
		Chemical ionization
Ash particle [48]	Bridging the gap, and distorting the electric field	Constructing particle chains and short-circuiting the gap
		The superposition effect of the charged particle and background electric fields.

1. Flame Temperature

From a macroscopic perspective, according to the ideal gas law, under constant pressure conditions, the high temperature of the flame continuously heats the gas in the gap, causing the gas volume to expand and the relative density of the gas to decrease, which is closely related to the breakdown voltage of the gap. You Fei et al. [49] studied the breakdown characteristics of a cedar pile fire under high-voltage power frequency, comparing the effects of high temperature and the addition of KCl to promote ionization on the breakdown voltage. They found that the decrease in air density due to the high temperature was the main reason for the reduction in gap insulation strength. Zhou Enze et al. [6], based on wildfire simulation breakdown test data at different altitudes, found that the effect of air density on flame gap insulation strength decreases with increasing altitude. The air density of the gap is also affected by the environmental altitude.

From a microscopic perspective, the high temperature of the flame affects the chemical reaction rate during vegetation combustion, the number of charged particles, and the development rate of the discharge in the arc. Feng Jing et al. developed a plasma microscopic numerical simulation model based on chemical–fluid dynamics to study the effect of flame gap temperatures of 293.15 K, 473.15 K, 673.15 K, and 873.15 K on discharge development. They found that at the exact moment when the maximum temperature was 873.15 K, the high electron density region advanced to a distance of 4 mm from the sample. In comparison, at 293.15 K, the high-electron-density region only advanced to 1.8 mm from the anode. Further calculations of discharge development rates revealed that at 873.15 K, the average development rate of the discharge could reach 0.427 mm/ns, which is an increase of 39% compared to normal temperature conditions [47].

Under the influence of a high flame temperature, the decrease in air density and the increased ionization of microscopic, charged particles promote the development rate of discharge in the gap, ultimately forming a discharge arc that threatens the insulation performance of the transmission line gap.

2. Conductivity

Many charged particles are generated during vegetation combustion through ionization and chemical reactions. These particles are influenced by flame convection and electric fields and enter the flame channel. This is macroscopically reflected by flame conductivity. These positive and negative ions converge into the discharge channel, providing the conditions for penetration.

In simulation experiments, the leakage current is an essential physical quantity reflecting flame conductivity. Huang Daochun et al. [30] observed the leakage current waveforms of three types of vegetation flames: straw, reed, and fir branches. The combustion-related chemical reactions in fir branches were the most intense, with a leakage current peak value of 27 mA, which is much higher than the other two vegetation types, and its breakdown voltage was also the lowest. Lu Wei et al. [7] studied the leakage current waveforms of multiple high-risk vegetation flames and found that their changes were consistent with flame morphology. This indicates that flame conductivity is almost synchronized with the intensity of combustion. At the peak of the fire, it has high conductivity, making it highly susceptible to gap discharge.

3. Smoke Particles

Due to the incomplete combustion of forest vegetation, the flame gap is filled with ash and particles. On the one hand, these particles can form particle chains that short-circuit the gap, increasing the risk of discharge during the non-bridging phase of the flame; on the other hand, the presence of particles distorts the electric field, forming localized substantial spatial charges, which promote the development of the discharge channel, eventually leading to gap discharge breakdown.

Yao Wenjun et al. from the Huazhong University of Science and Technology described the development process of the discharge channel in the presence of particles in the gap. The front of the discharge channel approaches the particles, and under the influence of the electric field at its head, the polarization of the particles is enhanced. The interaction with charged particles in the field causes the surrounding environmental electric field to distort. These particles also occupy space for electron avalanche development, revealing a competitive relationship between the two. The development of the discharge channel also depends on the particle size, with the channel containing particles of the same size or developing branches on the surface of larger particles [46].

The mechanism analysis based on simulation experiments primarily relies on theoretical models and experimental data to reveal the intrinsic coupling relationships between flame temperature, conductivity, smoke particles, and other environmental factors and physicochemical phenomena. However, due to the limitations of experimental conditions, there is still significant difficulty in conducting mechanism analysis for physical quantities that cannot or are difficult to be directly measured by experimental methods.

3.3.2. Mechanism Analysis Based on Numerical Simulation

The discharge channel is regarded as low-temperature plasma and is the main process of gap discharge. The development of gap discharge to break down under wildfire conditions is accompanied by various physical phenomena such as multiphase heat transfer, mass transfer, photoelectric effects, and complex chemical reactions, making it a typical multi-physics coupling problem. Its coupling relationships are shown in Figure 2.

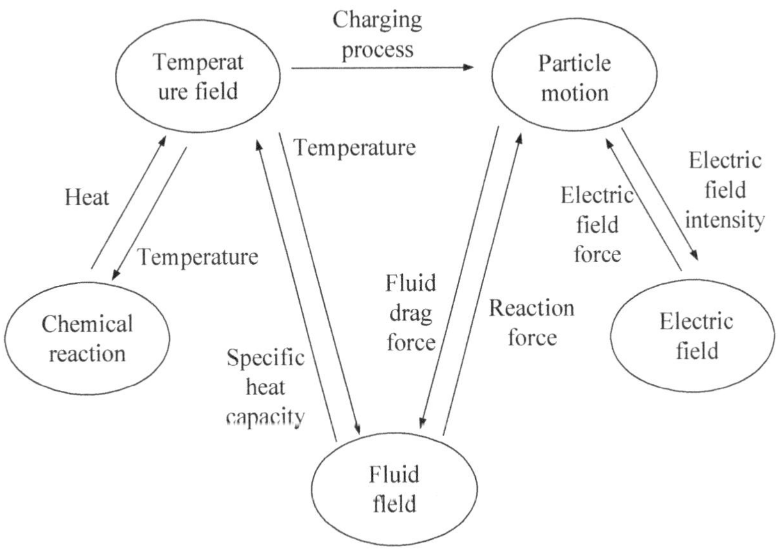

Figure 2. Multi-physics field coupling relationship.

Three types of plasma discharge models have been proposed to address the coupling of multiple physical fields: the Particle-in-Cell/Monte Carlo Collision (PIC/MCC) model, the fluid dynamics model, and the hybrid model.

1. Particle-in-Cell/Monte Carlo Collision (PIC/MCC) Model

This model is a key theoretical tool for studying streamers and other plasmas. It combines Particle-in-Cell (PIC) technology with Monte Carlo Collisions (MCCs) to simulate discharge streamers and can describe the motion of individual simulated electrons in the discharge process in real time. Y. Wang et al. conducted numerical simulations using a three-dimensional Particle-in-Cell/Monte Carlo Collision (PIC-MCC) model. Based on the assumptions of the field enhancement factor and field emission mechanism, they systematically analyzed the nanosecond pulsed breakdown characteristics of nitrogen spark switches at atmospheric pressure [50]. Numerical investigations of the plasma system were carried out by Z. Chen et al. using a 2D3V (two-dimensional, three-velocity-component) Particle-in-Cell model integrated with Monte Carlo Collision processes (PIC-MCCs). The temporal sampling of plasma and electromagnetic field distributions enabled the extraction of detailed information regarding the internal evolution of the plasma jet and the localized enhancement of electric fields [51].

However, this model has a significant drawback: it requires substantial computational resources and has a long convergence time, making it difficult to apply widely [52,53].

2. Fluid Dynamics Model

Baisen Lin and others from the South China University of Technology established a multi-physics coupling simulation model involving the electric field, thermal field, fluid field, chemical field, and particle motion to analyze the breakdown mechanism of air gaps under AC and DC conditions influenced by smoke particles. Compared to DC conductors, the percentage of smoke components entering the AC conductor detection area was 34.1% higher, and the smoke bridging percentage was 45% higher, increasing the likelihood of gap breakdown. This study examined the inherent mechanism of air gap electric field distortion at the microscopic level, providing theoretical support for revealing the breakdown process of transmission line gaps under wildfire conditions [54].

Jiang Wenquan from Hunan University established an arc discharge model to study the effects of high temperature and particulate matter on gap discharge. A two-dimensional axisymmetric model was chosen for the simulation. The data provided by the LXcat database [55] were directly used in the study. The discharge characteristics of the gap were studied under the effects of high temperature, particles, and the combination of high temperature–particle influence.

Pu Ziheng and others simulated the temperature distribution of the straw flame gap, the force analysis of burning particles, and the motion distribution based on existing experimental data [56]. A two-dimensional axisymmetric model, as shown in Figure 3, was adopted to simplify the calculations and facilitate convergence. The vegetation stack and the flame body were simplified as a cylinder and a frustum of a cone, respectively. The microscopic reasons for the particles moving away from and being absorbed by the electrodes were explained, and a criterion for determining whether the electrodes absorb particles was proposed by calculating the critical charge-to-mass ratio. The chemical reaction equations were further considered, simulating the effects of positive and negative polarity DC voltages on the flame morphology and particle concentration during the combustion of the wood stack. The intrinsic causes of positive and negative polarity DC voltages affect the temperature, height, and morphological changes in the vegetation flame, and the concentration of charged particles was revealed from a microscopic mechanism perspective [57].

Wang Ziming and others developed the saturation charge and charge-to-mass ratio of graphite particles with three different mesh sizes based on the particle charge-to-mass ratio and established a coupled simulation model for the temperature field, fluid field, particle dynamics field, and electric field. They studied the particle distribution patterns and electric field distortion characteristics under different charge states and particle sizes, as well as the synthetic field strength distribution curves of a single-flame "unimodal" (without considering space charge density) and "bimodal" (with space charge density) model. They also analyzed the discharge characteristics with the lowest insulation strength in the gap during a single oscillation cycle. Their flame turbulence model simultaneously considers the effects of a low Reynolds number flow and a vortex correction model, which can represent the phenomenon of vortex flames in the flame body [12].

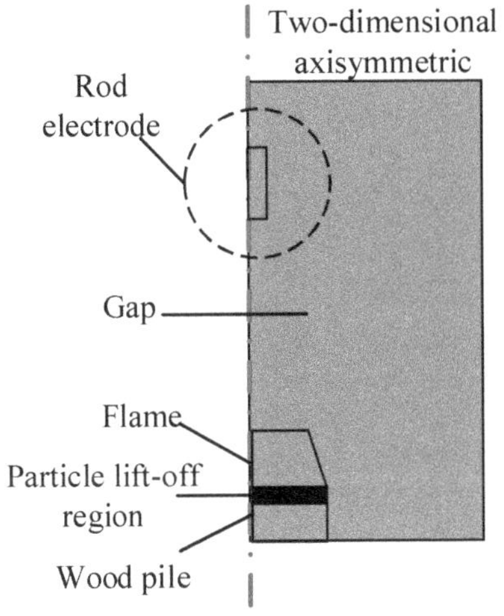

Figure 3. Schematic diagram of the simulation model.

However, multiple simplifications were made during the model construction process, such as assuming particles that have saturated charge, neglecting differences in particle shapes, and treating them all as spherical. Furthermore, the impact of factors such as wind speed and slope in the actual wildfire environment was not considered, so the simulation results needed further refinement.

3. Hybrid Model

The hybrid model is an improved model proposed by researchers based on the study of the stream discharge theory and the mechanism of gap discharge.

Farouk et al. proposed a fluid dynamics model that incorporates the Boltzmann electron equation, which combines the advantages of the particle/Monte Carlo model and traditional fluid dynamics models, achieving a balance between solution speed and computational accuracy [58]. Li and Ebert et al. divided the development space of the flow discharge into regions. In the flow channel area with intense ionization and low-field strength, they used a fluid dynamics model to describe the process, while in the region where the flow discharge was not reached, with lower electron density, the Monte Carlo model was more appropriate. At the boundary between these two regions, both models needed corrections to accurately describe the flow discharge development process [59].

However, the application cases of this hybrid model are limited, and its application environment is constrained, requiring further optimization in future research [60–63].

In conclusion, numerical simulation methods can partially compensate for the limitations of observing microscopic physical parameters in wildfire discharge simulation experiments. Analyzing parameters such as changes in charged particle concentrations, electric field distributions, and the motion trajectories and distribution patterns of particles in the discharge gap helps reveal the mechanisms of gap discharge [64,65]. Among them, numerical simulation methods using fluid dynamics models as the modeling approach are relatively mature and widely applied. However, existing research mostly uses rod-to-plate or sphere-to-plate two-dimensional axisymmetric models to simulate the uneven field discharge environment of actual power transmission lines to the ground. In the current advancement of large-scale gap discharge experiments, there is a lack of numerical simulation validation for long gap scales [66]. There is insufficient research on establishing numerical simulation models that closely reflect the actual transmission line corridor environment and meet the requirements for convergence and accuracy, as well as on the mechanisms of the inherent coupling effects of factors such as temperature, charged particle concentration, and particles [67].

4. Discussion

In summary, domestic and international research has extensively studied the breakdown characteristics of air gaps in transmission lines under wildfire conditions and developed gap breakdown voltage prediction models suitable for various wildfire scenarios. However, the current research on gap breakdown characteristics under wildfire conditions primarily relies on breakdown simulation experiments with limited experimental conditions. Domestic and international studies lack experimental data for gap distances of 9 m or more, which hinders their ability to provide references for large-scale gap line breakdown risk assessments under complex terrain and meteorological wildfire conditions.

Therefore, considering the need for improved temperature measurement methods that can capture discharge arc parameters and simulate the complex geographical conditions of actual transmission lines, the future development of a simulation experiment platform that closely aligns with the actual transmission line corridor environment, incorporating multi-parameter measurements and conducting in-depth numerical simulation considering climate and terrain factors, is envisioned.

1. The theoretical models with which to determine various influencing factors are independently scattered. The breakdown of transmission line gaps under wildfire conditions results from the coupling of multiple influencing factors. Domestic and international studies often consider single influencing factors. It is recommended that the coupled mathematical relationships of multiple wildfire characteristic parameters be studied, a discharge mathematical model for transmission lines under wildfire conditions based on multi-factor interactions be established, and measured data for correction be used to improve the accuracy of breakdown voltage predictions.

2. There are few simulation experiments on the long-gap AC and DC discharge characteristics of actual transmission lines under flame conditions. The review summarizes many small-scale simulation experiments, mostly at the centimeter level and below 4 m. Establishing a long-gap, multi-parameter real-time monitoring, and multi-variable, effectively controllable composite functional simulation experiment platform is urgently needed. This platform can conduct simulation tests closer to the actual transmission line corridor environment, providing practical reference data and conclusions for guiding insulation failure risk assessments and dynamic defense work for ultra-high voltage transmission lines under wildfire conditions.

3. Research on arc morphology and discharge parameters during the discharge process needs further investigation. Currently, high-speed cameras can be used to capture the macroscopic shape and direction of the arc during discharge, and multi-physics coupled fluid dynamics models can be constructed to simulate the changes in microparameters, such as local electric field strength, arc radius, and development rate, during the development of discharge streamers to analyze their mechanisms. However, there is still a lack of intuitive observational methods with which to capture the guiding effect of ash particles on the arc. Furthermore, numerical models constructed to meet the requirements of computational convergence and simplified analysis processes still cannot accurately simulate the complex and variable meteorological and topographical environments around actual transmission lines. For example, the effects of wind speed on arc path deviation and the influence of particle movement and distribution on guiding the arc path remain unclear, leaving significant room for further research.

5. Conclusions

1. Research on breakdown characteristics still mainly relies on conducting simulated gap breakdown experiments under wildfire conditions. Factors such as high flame temperature, high electrical conductivity, and dense smoke particles have always been the focus of experimental research. In addition, independent experiments have been conducted on various factors, including heat release rates and combustion calorific values, vegetation types, wind speed, and slopes. This also requires higher demand for the experimental platform's measurement system, which must be capable of handling multi-parameter, high-precision, and fast-response measurements. Comprehensive and accurate data collection from experiments is a prerequisite for reducing errors in the breakdown voltage prediction model. In simulated experiments, considering the coupling of multiple influencing factors and constructing a breakdown voltage prediction model with correction factors for each influencing factor is essential to effectively guide insulation failure risk assessment and dynamic defense measures for ultra-high voltage transmission lines under wildfire conditions.

2. The mechanism analysis based on simulated experiments primarily relies on three fundamental theoretical models: the decrease in air density, high flame conductivity, and particle-triggered discharge, combined with simulation experiment data to analyze the discharge mechanism of flame gaps. Progress has been made in studies on coupling decreases in air density with altitude factors, coupling high-flame conductivity with chemical reactions, and the interaction between smoke particles and streamer discharge. However, due to limitations in experimental conditions, many microscopic physical quantities are difficult to observe directly, hindering the ability to conduct more in-depth mechanism analysis.

3. Mechanism analysis based on numerical simulation transforms flame gap discharge into a multi-physics coupling problem, constructing simulation models to study the microscopic physical parameters during the discharge process, which, to some extent, compensates for the limitations of observing microscopic physical parameters in wildfire discharge simulations. Among these, the numerical simulation method based on fluid dynamics models is relatively mature and widely applied, which may be the most accurate and applicable model for gap discharging simulations under mountain fires, especially considering temperature, conductivity, and ash. However, there is a lack of numerical simulation validation for long gap breakdown experiments. There is insufficient research on the inherent coupling mechanisms of temperature, charged particle concentrations, and particle matter when establishing a simulation model closely resembling the actual transmission line corridor environment.

Funding: This research was funded by the National Natural Science Foundation of China, grant number 51677138.

Conflicts of Interest: The authors declare no conflict of interest.

References

1. He, C.; Shu, L. Study on Environmental Characteristics of High Voltage Line Trip Caused by Mountain Fire in Guangxi Region. *Fire Sci. Technol.* **2020**, *39*, 1626–1629.
2. Zhou, E.; Gong, B. Statistics and Analysis on Tripping Fault of Overhead Transmission Lines Caused by Mountain Fires in China Southern Power Grids. *Guangdong Electr. Power* **2022**, *35*, 80–86.
3. Zhou, Z. *Research on Risk Assessment of Power Grid Transmission-Line Trips Due to Wildfire Disasters*; North China Electric Power University: Beijing, China, 2019. [CrossRef]
4. Li, P.; Ruan, J. Influence Factors Analysis of the Conductor-Plane Gap Breakdown Characteristic Under Simulation Forest Fire Condition. *Trans. China Electrotech. Soc.* **2018**, *33*, 195–201. [CrossRef]
5. Long, M.; Huang, D. Influence of the Typical Vegetation Ashes/Particles on the Discharge Characteristics of Conductor-Plane Air Gap. *Trans. China Electrotech. Soc.* **2018**, *33*, 627–633. [CrossRef]

6. Zhou, E.; Fan, L. Breakdown Characteristics of Conductor-plane Gap Under Vegetation Fire at the Altitude of 2013 m. *High Volt. Eng.* **2022**, *48*, 4316–4322. [CrossRef]
7. Lu, W. Simulated Experimental Study on the Breakdown Characteristics of the Gap with High-Risk Vegetation Flame. Master's Thesis, Wuhan University, Wuhan, China, May 2019.
8. Wu, M. The Evolution of Gas Laser Al Plasma in the Background Interference Diagnosis Research. Master's Thesis, Northwest Normal University, Lanzhou, China, May 2023. [CrossRef]
9. Yang, N.; Yang, B. Experimental Research on Impact of Mountain Fire Smoke on Insulation of 10~35 kV Transmission Lines. *Sichuan Electr. Power Technol.* **2022**, *45*, 6–9. [CrossRef]
10. Hu, C. Research on External Insulation Characteristics And Laser Lidar Monitoring System of Forest Fire in Transmission Line. Master's Thesis, South China University of Technology, Guangzhou, China, 2016.
11. Liu, X. Study on Smoke Concentration Measurement by Pulse Laser. Master's Thesis, Beijing Institute of Technology, Beijing, China, January 2016.
12. Wang, Z. Research on the Distribution Law of Charged Particles in the Insulation Gap and the Characteristic of Synthetic Electric Field Distortion under Mountain Fire. Master's Thesis, China Three Gorges University, Yichang, China, May 2022.
13. Ma, Q.; Zhang, R.; Wang, W. Research on Effect of Single Thermocouple's Placement on Temperature Calibration for Fire Test. In Proceedings of the 2020 IEEE 2nd International Conference on Civil Aviation Safety and Information Technology, Weihai, China, 14–16 October 2020; pp. 996–1003. [CrossRef]
14. Chen, Y.; Zhang, H. Two-dimensional temperature field measurement by laser interference based on fringes Fourier transformation. *J. Univ. Sci. Technol. China* **2020**, *50*, 832–838.
15. Sun, Z. Research on Flame Temperature Based on Laser Interferometry. Master's Thesis, North China Electric Power University, Beijing, China, April 2020. [CrossRef]
16. Lin, T.; Deng, Y. Flame Temperature Measurement Based on Atomic Spectroscopy. *J. Instrum. Anal.* **2021**, *40*, 785–789.
17. Jiang, W. Impacts of Forest Fire on Streamer Discharge of Transmission Line GAP. Master's Thesis, Hunan University, Changsha, China, April 2019. [CrossRef]
18. Grange, F.; Soulem, N. Numerical and Experimental Determination of Ionizing Front Velocity in a DC Point-to-Plane Corona Discharge. *J. Phys. D Appl. Phys.* **1995**, *28*, 1619.
19. Sigmond, R.S. The Residual Streamer Channel: Return Strokes and Secondary Streamers. *J. Appl. Phys.* **1984**, *56*, 1355–1370.
20. He, H.; He, J. On the Positive upward Leader in Response to Downward Stepped Leader in a 10 m Rod-to-Rod Long Air Gap. In Proceedings of the International Conference on Lightning Protection, New York, NY, USA, 25–30 September 2016. [CrossRef]
21. Hidaka, K.; Fujita, H. A New Method of Electric Field Measurements in Corona Discharge Using Pockels Device. *J. Appl. Phys.* **1982**, *53*, 5999–6003. [CrossRef]
22. Zeng, R.; Zhuang, C. Electric Field Step in Air Gap Streamer Discharges. *Appl. Phys. Lett.* **2011**, *99*, 221503. [CrossRef]
23. Zeng, R.; Wang, B. Design and Application of an Integrated Electro-optic Sensor for Intensive Electric Field Measurement. *IEEE Trans. Dielectr. Electr. Insul.* **2011**, *18*, 312–319. [CrossRef]
24. Chun, K.; Song, D.; Kim, K. Study on the Decrease in Air Dielectric Strength of DC Voltage According to the Ignition Properties of Combustibles. *Appl. Sci.* **2022**, *12*, 2216. [CrossRef]
25. Ntshangase, Z.; Swanson, A.; Ijumba, N.M.; Britten, A.C. An Empirical Model of Fire-Induced Airgap Voltage Breakdown Characteristics Under HVDC Conditions. In Proceedings of the 2017 IEEE AFRICON: Science, Technology, and Innovation for Africa, Cape Town, South Africa, 18–20 September 2017; pp. 1283–1289.
26. Robledo-Martinez, A.; Guzman, E.; Hernandez, J.L. Dielectric Characteristics of a Model Transmission Line in the Presence of Fire. *IEEE Trans. Electr. Insul.* **1991**, *26*, 776–782. [CrossRef]
27. Fonseca, J.R.; Tan, A.L. Effects of Agricultural Fires on the Performance of Overhead Transmission Lines. *IEEE Trans. Power Deliv.* **1990**, *5*, 687–694. [CrossRef]
28. Moreno, M. *Performance of External Insulation in Presence of Flames Due to Sugar Cane Burning*; CIGRE: Paris, France, 1985; CIGRE WG 33-07/10 IWD.
29. Lanoie, R.; Mercure, H. Influence of Forest Fires on Power Line Insulation. In Proceedings of the 6th International Symposium on High Voltage Engineering (ISH), Xi'an, China, 21–26 November 2021; pp. 12–18.
30. Huang, D.; Lu, W.; Yao, T.; Xia, J.; Quan, W. Study on Leakage Current Characteristics of Conductor-Plane Short Air Gap Under Vegetation Fire Condition. *J. Electr. Eng. Technol.* **2019**, *34*, 3487–3493. [CrossRef]
31. Huang, D.; Lu, W. Influence of Typical Vegetation Ashes on DC Breakdown Characteristics of Conductor-Plane Air Gap. *Power Syst. Technol.* **2018**, *42*, 1668–1674. [CrossRef]
32. Huang, D.; Chen, X. Breakdown Voltage of Conductor-Plane Gap Under Vegetation Fire Condition Considering Division of Flames. *Power Syst. Technol.* **2023**, *47*, 3467–3474. [CrossRef]
33. Huang, D.; Zhou, S. ACAC Breakdown Characteristics of Overhead Transmission Line Phase-Ground Long Air Gap Under Vegetation Fire Conditions. *Proc. CSEE* **2024**, *7*, 373. [CrossRef]

34. Li, P.; Ruan, J. Study on Breakdown Characteristic and Discharge Model of Conductor-Plane Gap Under Typical Vegetation Flame. *Proc. CSEE* **2016**, *36*, 4001–4011. [CrossRef]
35. Yang, K. Simulative Experimental and Numerical Study of Wildfire Induced Breakdown of Transmission Lines. Master's Thesis, University of Science and Technology of China, Hefei, China, May 2016.
36. Zhou, E.; Rao, Z.; Liu, Q.; Huang, Y.; Xiang, Z.; Wang, H.; Huang, D. Breakdown Characteristics of Conductor-Plane Gap Under Typical High Risk Vegetation Fire Condition. *High Volt. Appar.* **2024**, *60*, 179–185. [CrossRef]
37. Chen, X.; Wei, H. Study on the Influence of Flame Height on Gap Breakdown Characteristics of Transmission Lines. *Electr. Power* **2015**, *48*, 134–138.
38. Wang, T.; Zhou, E. Influence Factor Analysis of Conductor-Plane Gap Breakdown Characteristic Under Vegetation Fire at High Altitude Condition. *Guangdong Electr. Power* **2024**, *37*, 70–76.
39. Liu, H.; He, Z. Research of electric field characteristics of high-voltage overhead transmission lines adjacent to trees. *Zhejiang Electr. Power* **2023**, *42*, 70–78. [CrossRef]
40. Wang, D.; Lu, T. Modelling and Characteristic Research of Total Electric Field Around Trees Near ±1100 kV UHVDC Transmission Lines. *Power Syst. Technol.* **2017**, *41*, 3441–3447. [CrossRef]
41. Shu, S.; Liu, C.; Ruan, J. Review on Prediction Methods for Breakdown Voltage of Air Gap. *High Volt. Appar.* **2016**, *52*, 19–26+32. [CrossRef]
42. Khan, I.; Ghassemi, M. A Probabilistic Approach for Analysis of Line Outage Risk Caused by Wildfires. *Int. J. Electr. Power Energy Syst.* **2022**, *139*, 108042. [CrossRef]
43. Song, J.; Guo, C.; Zhang, J.; Wang, Y.; Sheng, K. A Probabilistic Model of Overhead Transmission Line Outage due to Forest Fire. *Power Syst. Technol.* **2013**, *37*, 100–105. [CrossRef]
44. Shao, Y.; Li, H.; Zhao, J.; Zou, J.; Zhou, X.; Chang, Q. Study and Analysis of Breakdown Probability Model of UHV Transmission Line under Forest Fire Conditions. *Electr. Meas. Instrum.* **2018**, *55*, 31–35+42.
45. Zhou, C. Numerical Simulation Study on the Influence of Wildfire on Gap Discharge of Transmission Line. Master's Thesis, Chongqing University, Chongqing, China, June 2021. [CrossRef]
46. Yao, W. The Research on the Characteristics of the Two-Phase Mixture Discharges under the DC Voltage. Ph.D. Thesis, Huazhong University of Science and Technology, Wuhan, China, November 2013.
47. Feng, J. Study on Gap Discharge Mechanism of Power Transmission and Distribution Lines Under Different Wildfire Conditions. Master's Thesis, Xihua University, Chengdu, China, April 2023. [CrossRef]
48. Huang, D.; Li, P. Review on Discharge Mechanism and Breakdown Characteristics of Transmission Line Gap Under Forest Fire Condition. *High Volt. Eng.* **2015**, *41*, 622–632. [CrossRef]
49. You, F.; Chen, H. Experimental Study on Flashover of High-voltage Transmission Lines Induced by Wood Crib Fire. *J. Proc. CSEE* **2011**, *31*, 192–197. [CrossRef]
50. Wang, Y.; Li, J.; Li, C. The effects of field-enhancement factor on the nanosecond pulsed breakdown of nitrogen spark switch at atmospheric pressure with 3D PIC-MCC model. In Proceedings of the 2024 IEEE International Conference on Plasma Science (ICOPS), Beijing, China, 16–20 June 2024; p. 1. [CrossRef]
51. Chen, Z.; Liu, M.; Xia, G.; Huang, Y. PIC/MCC Simulation of the Ionization Process for Filamentary Streamer Plasma Jet at Atmosphere Pressure in Argon. *IEEE Trans. Plasma Sci.* **2012**, *40*, 2861–2865. [CrossRef]
52. Babaeva, N.Y.; Kushner, M.J. Effect of Inhomogeneities on Streamer Propagation: I. Intersection with Isolated Bubbles and Particles. *Plasma Sources Sci. Technol.* **2009**, *35*, 035009.
53. Peng, Q. Research of Plasma Chemical Model and Analysis of Influencing Factors of Streamer Discharge in Air. Ph.D. Thesis, Chongqing University, Chongqing, China, October 2012.
54. Lin, B. Simulation Analysis on the Synergistic Effect of Vegetation Ashes and Charged Particles on the Gap Electric Field Distortion. *J. Phys. D Appl. Phys.* **2024**, *57*, 205501. [CrossRef]
55. Hagelaar, G.J.M.; Pitchford, L.C. Solving the Boltzmann Equation to Obtain Electron Transport Coefficients and Rate Coefficients for Fluid Models. *Plasma Sources Sci. Technol.* **2005**, *14*, 722–733.
56. Pu, Z.; Wang, Z. Motion and Distribution of Combustion Particles in Flame Gap Under DC Voltage. *Trans. China Electrotech. Soc.* **2020**, *35*, 612–619. [CrossRef]
57. Pu, Z.; Sun, R. Mechanism Analysis of Effect of Electric Field on Vegetation Flame Under DC Voltage. *J. Combust. Sci. Technol.* **2022**, *28*, 379–388.
58. Farouk, T.; Farouk, B. Simulation of DC Atmospheric Pressure Argon Microglow-Discharge. *Plasma Sources Sci. Technol.* **2006**, *15*, 676–688.
59. Li, C.; Ebert, U.; Hundsorfer, W. Spatially Hybrid Computations for Streamer Discharges with Generic Features of Pulled Fronts: I.Planar Fronts. *J. Comput. Phys.* **2010**, *229*, 200–220. [CrossRef]
60. Yang, J.; Ren, X.; Wang, X.; Zhang, X.; Fan, S.; Chen, T. Experimental Research on Influence of Mountain Fire on Safe Operation of 10~35 k V Transmission Lines. *Sichuan Electr. Power Technol.* **2022**, *45*, 1–5. [CrossRef]

61. Yuan, T.; Yang, Z. Study on Impluse Flashover Arc Observation and Post-Arc Gas Dissipation Process in the Semienclosed Chamber. *Trans. China Electrotech. Soc.* **2024**, *39*, 924–934. [CrossRef]
62. Liu, S. Observation of Vacuum Arc Morphology and Microscopic Particle Dynamics. Master's Thesis, Huazhong University of Science and Technology, Wuhan, China, May 2021. [CrossRef]
63. Yang, C.; Ning, X.; Xu, H.; Chen, T.; Zeng, C.; Yang, S. Overview of Mountain Fire Monitoring and Early Warning for Power Grid Demand. *Power Syst. Technol.* **2023**, *47*, 4765–4777. [CrossRef]
64. Zhang, K.; Wu, X. Real-Time Wildfire Risk Assessment Model for Transmission Corridors Based on Feature Engineering and Boosting Integrated Learning Model, Ensemble Learning, and Model Fusion. *Power Syst. Technol.* **2023**, *47*, 4727–4738.
65. Song, H.; Meng, X.; Sheng, G.; Jiang, X. Overview of Experimental Observation Technology for Short Air Gap Streamer Discharge. *Power Syst. Technol.* **2022**, *46*, 774–785. [CrossRef]
66. Wang, P.; Liu, X. Stepped Propagation Characteristics of Leader for Positive Discharge in Long Air Gaps. *Power Syst. Technol.* **2021**, *45*, 818–824. [CrossRef]
67. Zhou, E.; Fan, L. Risk Distribution Assessment of Wildfire-Induced Trips in Transmission Line Based on Flame Combustion Model. *Power Syst. Technol.* **2022**, *46*, 2778–2785. [CrossRef]

energies

Article

Corona-Generated Space Charge Characteristic in an Indoor HVDC Corona Cage Under Atmospheric Temperature Conditions

Jules Simplice Djeumen *, Hendrick Musawenkosi Langa and Trudy Sutherland *

Department of Electrical Engineering, Faculty of Engineering and Technology, Vaal University of Technology, Vanderbijlpark 1900, South Africa; hendrickl@vut.ac.za

* Correspondence: julesd@vut.ac.za (J.S.D.); trudys@vut.ac.za (T.S.)

Abstract: This study conducted experiments and simulations to examine the DC corona-generated space charge characteristics and understand the performance of high-voltage direct current (HVDC) transmission lines. In experimental studies, various gradient temperatures are tested on a standard model of the potential HVDC transmission line in Southern Africa using an indoor corona cage. Initial tests on the single-line model of aluminium TERN conductors measured the DC corona inception voltages (CIVs) as the ambient temperature increased from 25 °C to 42 °C. A daylight ultraviolet corona camera (CoroCam8) has been used for measurements and visualisation; the measurements record temperatures for positive and negative direct current (DC) voltages. Experimental investigations are supplemented by simulations utilising the finite element method (FEM)-based software COMSOL Multiphysics. Following the creation of 3D models of the corona cage and potential conductor arrangement, the electric field distribution on the surfaces of the conductors was examined. The CIV observations and modelling findings determine the setups' corona inception electric field strengths. The study effectively integrated experimental data from a corona cage with FEM models to assess DC corona properties across different air temperatures thoroughly. The inception voltage levels of corona are significantly influenced by ambient temperature and the space charge generated by corona. The outcomes of the discussion will inform the design of the proposed HVDC transmission line in Southern Africa.

Keywords: corona discharge; corona inception voltage; electric field; space charge; corona cage

1. Introduction

High-voltage direct current (HVDC) transmission has grown in demand as the amount of electricity transmitted over long distances has increased significantly. As a result, numerous HVDC transmission lines have been constructed globally [1,2]. HVDC transmission technology offers technical, financial, and environmental advantages over the standard AC grid when transmitting bulk electrical energy over long distances or asynchronous interconnections [3,4]. The need to expand transmission capacity has grown with the steady increase in power demand. Governments face numerous challenges in meeting the increasing demand for electricity due to constraints, such as voltage range, conductor limitations, high costs, and limited land availability for transmission line tower design. Even though HVDC has many benefits, such as fewer stability problems, no synchronisation concerns, lower prices, etc., there always needs to be adequate backing for the new power

 https://doi.org/10.3390/en18112872

lines. In addition to its well-known advantages, HVDC systems exhibit greater corona performance than conventional AC transmission. HVDC transmission is anticipated to be increasingly significant in future power systems, introducing novel issues for electrical engineers. A self-sustained corona discharge transpires in the atmospheric vicinity of a conductor when the voltage gradient on its surface exceeds a defined threshold. Corona discharge is affected by environmental factors, including humidity and temperature, as well as the nature of the voltage, conductor material, conductor diameter, and surface roughness [3]. The corona properties of transmission lines considerably influence their design and affect line performance [5,6]. Corona-induced space charge, corona inception voltages (CIVs), and corona losses (CL) are among the most significant aspects to consider in the design of transmission systems. The conductor size and ambient conditions influence these parameters [7].

Computational models are essential for analysing and forecasting the corona effects in HVDC lines linked to HVAC systems. These models are especially beneficial for examining the intricate interplay between AC and DC fields and their influence on space charge properties. Simulations based on the finite element method (FEM) are extensively employed to study electric field distribution and space charge dynamics in corona cages. Experimental data validate these simulations to ensure precision [8,9]. The time-domain finite element method is utilised to analyse transient overvoltages and their impact on corona discharge. It offers insights into the dynamic behaviour of space charges under different electric field situations [10]. The hybrid simulation technique integrates the finite element technique (FEM) and finite volume methods to simulate ion flow fields and corona currents in hybrid AC-DC systems. These techniques effectively examine corona losses and ion current distributions [11,12]. Corona characteristics may be computed using empirical formulas and simulations of the field distribution [13,14] or may be discovered through experimental research. Calculations based on simulations might produce erroneous conclusions due to the complexity of the corona phenomenon. Conversely, experimental research employing single-phase overhead test lines is constrained by their high costs and insufficient laboratory space [15]. To effectively assess the corona properties of the surface's transmission line, using corona cages and testing under varying temperature conditions is recommended due to their efficiency, speed, and cost-effectiveness [16]. To replicate the operational environment accurately, the cage's dimensions and the conductors' lengths must be adequate. For the specified voltage level and conductor geometry, it is recommended that the cage be designed utilising simulations.

The determination of CIV has been extensively investigated through various experimental and numerical studies documented in the literature. The assessment of corona losses, electric field strengths, and current density profiles associated with single monopolar HVDC transmission lines was detailed in references [17,18], utilising a FEM-based methodology; nevertheless, the influence of temperature on these phenomena is not well-defined in the existing literature. Semi-empirical formulations were developed for fair and hot weather conditions to calculate corona losses in monopolar HVDC lines by interpolating experimental data [19]. However, the reliability and stability of HVDC transmissions depend on a clear understanding of the physical interactions between the transmission environment and the electrical infrastructure. One of the most critical phenomena affecting HVDC performance is the corona discharge, which forms space charges in the surrounding air around the transmission conductors. These space charges modify the electric field distribution, potentially leading to conductor stress, partial discharges, and long-term material degradation.

Atmospheric temperature plays a crucial role in this process [20,21]; it influences the corona onset voltage by affecting air density, ion mobility, and space charge drift within and around the HVDC line conductors [22,23]. Understanding these temperature-related effects is crucial for optimising HVDC line design, enhancing system reliability, and minimising environmental impacts, such as corona inception voltage, radio noise, and audible sound [24,25].

2. Experimental Setup

2.1. Corona Cage 3D Design

The experimental laboratories began with software design, where the real corona cage was first designed and tested before its construction. Several corona cage designs have been developed, and the final design is presented in Figure 1 below. A laboratory corona cage serves as an instrument for assessing the corona performance of conductors. It consists of a concentric metal cage with a substantial radius, available in either a circular or square cross-section. It incorporates a test line at its centre to replicate a single or multiple (bundle) conductor(s). It was decided to design a cage with a diameter of 1.5 m, which will be housed at the Vaal University of Technology's HV Laboratory. This diameter would ensure that corona studies can be carried out without fearing flashovers.

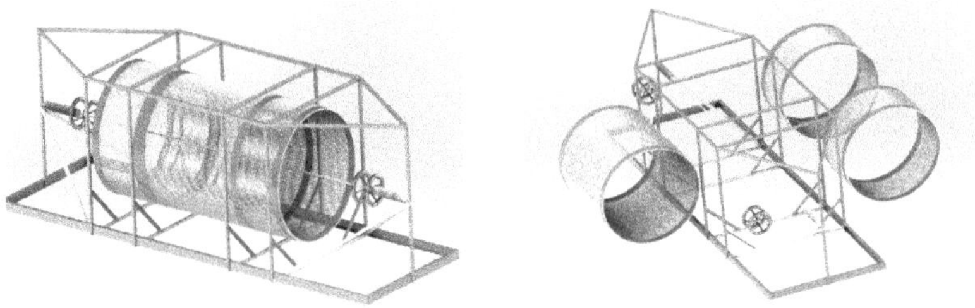

(a) (b)

Figure 1. (**a**) Three-segment cage, (**b**) final cage structure designed.

The cage's structure is divided into the following three distinct sections: two outer guard rings, each measuring 0.5 m long, and one central measuring section, which is 1 m long. A support structure was built to support the cage and be meshed for a proper simulation, as shown in Figure 1b.

A corona cage is a single-phase testing facility, where conductors or conductor bundles are positioned centrally within a grounded mesh cage. The primary purpose of the corona cage configurations and dimensions is to accurately replicate the electric field distribution in the specified region surrounding the conductor through a small-scale measurement setup [26].

A corona cage is an ideal tool for studying the corona-generated space charge performance of transmission lines and conductors in a laboratory setup. Its practicability has been confirmed by [9,17–20,26–28], which presents several advantages; for example, it reduces the test space, as shown in Figure 2, has low investment, controllable test conditions, a convenient structure for adjustment, and a short test period compared to other test methods mentioned below.

Many studies have been conducted on the circular corona cage, which consists of the following three parts: a middle part for measurement and two sides for protection. Different corona cages have been employed for the corona test studies, as follows: the small or indoor corona cage [17,18], the outdoor or larger corona cage, the mobile corona cage, and the outdoor test line technique [9,19,20,26–29]. Although the test line remains the most efficient, its results approach the characteristics of practical transmission lines; this approach is very costly and hazardous.

Figure 2. Virtual experimental setup in HV Laboratory.

CIV is the lowest voltage at which the continuous corona of specified pulse amplitude occurs as the applied voltage gradually increases. The CIV decreases as the frequency of the applied voltage increases. Under HVDC conditions, the inception voltage for negative corona on conductors is observed to be lower than that of the positive case, generally approximating 90% of the positive corona inception voltage [30,31].

2.2. Real Experimental Arrangement in the HV Laboratory

An indoor corona cage, measuring 387 cm in length and 150 cm in diameter, was used in the experiments. It was designed for corona discharge measurements and features a cylindrical shape, consisting of two shield segments at both ends and a measuring segment in the middle that can be electrically connected. The shield segments are designed to be electrically isolated from the measuring segment(s) to facilitate the measurement of corona discharge occurring on the conductor surface, while simultaneously preventing corona discharge on the cage. Figure 3 presents the comprehensive setup schematic diagram of the corona cage.

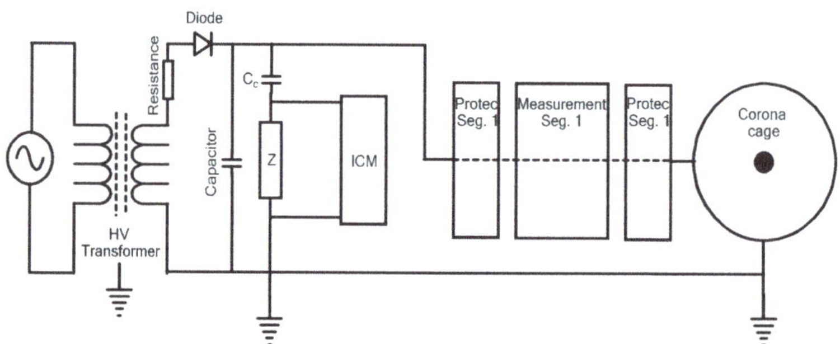

Figure 3. The system configuration.

The HVDC supply utilised in the experiments consists of several components, including a transformer, diodes, damping and limiting resistors, and doubling and smoothing capacitors capable of delivering a voltage of up to ±270 kV. The shield segments are electrically isolated from the measuring segment to facilitate the measurement of corona generated on the conductor surface and to prevent corona discharge on the cage [30].

2.3. Corona Inception Voltage at the Ambient Temperature (CIV)

The corona inception voltage represents the minimum voltage level at which continuous corona discharge of a specified pulse amplitude occurs as the applied voltage gradually increases. The inception voltage of corona is inversely related to the frequency of the applied voltage, exhibiting a decrease as the frequency increases. In HVDC conditions, the positive corona inception voltage for conductors is higher than the negative case and is typically about 10% higher than the negative CIV [13]. In the operation of a practical power system, aluminium conductor steel reinforced (ACSR), referred to by the code word "TERN", is commonly utilised instead of cylinder conductors for power transmission lines. The ACSR stranded conductors exhibit favourable electrical conductivity, sufficient mechanical strength, and tensile strength [7,8], the primary reasons for their use during the evaluation tests. Given the corona cage diameter and the conductor diameter, it is possible to estimate the CIV. This is often carried out using Peek's empirical constants (E_0 and K) in the well-known formula from the experimental data for the corona inception gradient E_c in kV/cm [15] in Equation (1) below for a smooth cylindrical conductor in air.

$$E_C = E_0\delta_0\left(1+\frac{0.308}{\sqrt{\delta_0 r_0}}\right) \quad (1)$$

where:

E_0: disruptive critical electric field of air ≈ 30 kV/cm (rms)
r_0: conductor radius (in cm)
δ_0: relative air density factor, resulting from:

$$\delta_0 = \left(\frac{3.92P}{273+T}\right) \quad (2)$$

where:

T: ambient temperature in °C
P: atmospheric pressure in mmHg

The inception voltage for negative corona for the HVDC conditions has been simplified using Peek's empirical constants (E_0 and K) with their average value [15].

For a conductor radius of 1 cm at 25 and 760 mmHg pressure, as an example:

$$E_C = 30 \times 1\left(1+\frac{0.308}{\sqrt{1\times 1}} = 30\times(1+0.308) \approx 39.24\ \mathrm{kV/cm}\right) \quad (3)$$

δ is given by Equation (4) below, namely:

$$\delta_0 = \left(\frac{3.92\times 760}{(273+25)}\right) \approx 1.0 \quad (4)$$

Figures 4 and 5 show the CIV visualisation via the ICM monitor, which displays the number of charges at the first corona discharge for the positive and negative voltages.

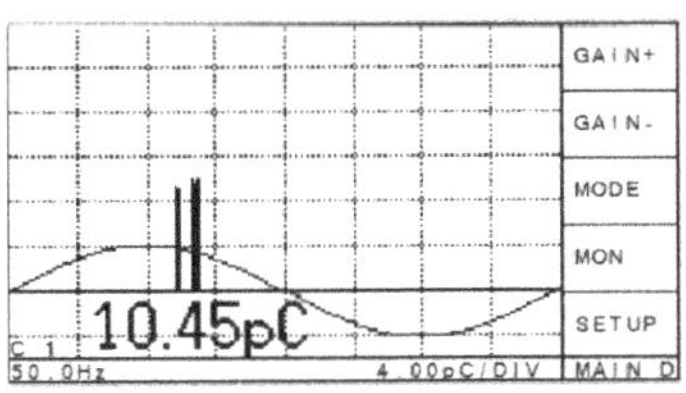

Figure 4. Positive inception voltage at 21 kV.

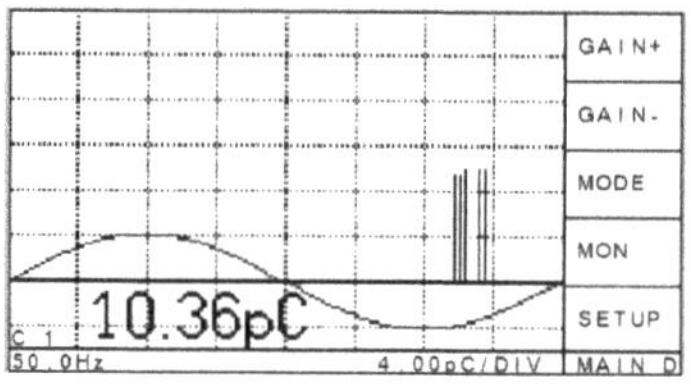

Figure 5. Negative inception voltage at −17 kV.

2.4. Environmental Temperature on the Corona Inception Voltage (Vinc)

The relationship between the corona inception voltage and temperature of the "TERN" conductor under the corona cage is summarised in Table 1.

Table 1. Temperature data summary.

T (°C)	25 °C	27 °C	29 °C	31 °C	33 °C	35 °C	37 °C	39 °C	41 °C	42 °C
Positive (+V_{inc})	+21 kV	+21 kV	+21 kV	+20.5 kV	+20.5 kV	+20 kV	+20 kV	+19.5 kV	+19.5 kV	+19 kV
Negative (−V_{inc})	−16 kV	−16 kV	−16 kV	−16.5 kV	−16.5 kV	−17 kV	−17 kV	−17.5 kV	−17.5 kV	−18 kV

Figures 6 and 7 below show the variation V_{inc} = f(T °C) function. Inception voltages range from −16 kV to −18 kV for negative voltage and +19 kV to +21 kV for positive voltage at temperatures between 25 °C and 42 °C, which correspond to the average ambient and maximum temperatures. It can be observed that the inception voltage decreases in both cases. This is mathematically true due to the presence of negative numbers. Still, it is not physically accurate, because the electric field is the cause of corona [9,27], and it is evident that the negative inception field decreases with increasing temperature.

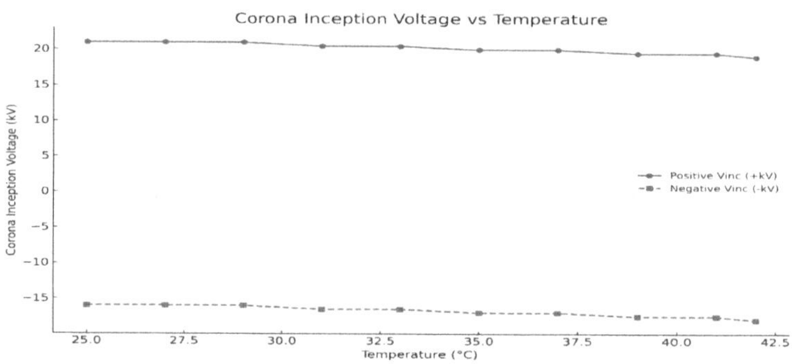

Figure 6. Relationship between corona inception and temperature for positive and negative voltage.

Figure 7 illustrates how corona inception voltage varies with temperature for positive and negative HVDC polarities. A gradual decline is observed in the positive inception voltage as the temperature increases, indicating reduced air dielectric strength. In contrast, the negative inception voltage drops more steeply, suggesting greater sensitivity of negative corona discharge to thermal effects. This polarity-dependent behaviour highlights the role of space charge dynamics and ion mobility under varying atmospheric conditions.

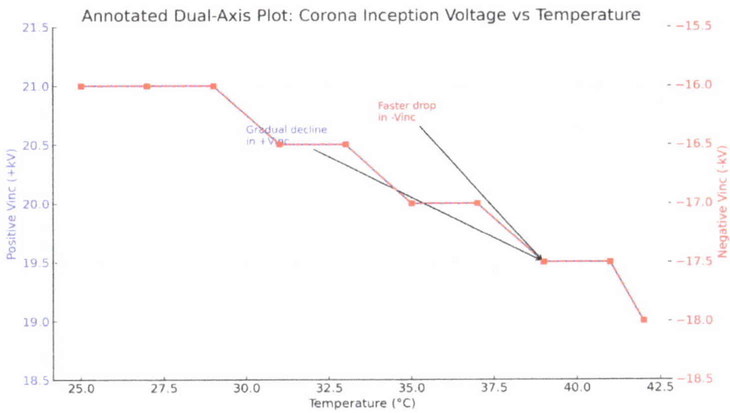

Figure 7. Annotated dual-axis plot showing the temperature-dependent behaviour of positive and negative corona inception voltages in an HVDC system.

2.5. Experimental Procedure in the High-Voltage Laboratory

Figure 8 illustrates the corona cage in the high-voltage laboratory at the Vaal University of Technology. The general indoor corona cage comprises the following three components: an inner section designated for measurement and two outer sections known as protective guards [20]. The two outer sections are identical and intended for electrical connection.

Figure 8. Test setup photograph with corocam8.

The conductor (TERN), the most used conductor in the transmission line, was placed along the axis of the corona cage and stretched via post insulators. The guard electrodes of the cage were connected to the ground, along with the measurement segment. Thick conductors and corona rings were implemented at the connection points to mitigate undesired discharges. A test voltage was applied to one side of the test conductors, with the voltage gradually increasing from the inception voltage to 175 kV. A corona ring was used to terminate the other side. The test arrangement photo is shown in Figure 9, featuring the two giant heaters that gradually increased the room temperature, with the voltage capped at its highest and lowest values at ±175 kV for both positive and negative voltages.

Figure 9. The corona cage with the heater on in the laboratory.

2.5.1. Effect of Temperature on Corona-Generated Space Charge for a Positive Voltage

The process of heating and monitoring the room was accomplished using two giant heaters, each with a capacity of 3 kW, and coupled parallelly with the thermos-anemometer environment meter tester model (MT948), which can measure air flow, temperature, and air velocity, and a digital temperature humidity meter with a measuring range of −10 to 50 °C for the environmental temperature for more accuracy. These heaters, the thermos-anemometer environment meter tester and the Fluke T3000 FC wireless digital thermometer, were used to gradually increase and monitor the room temperature, allowing it to rise from its normal ambient temperature of 25 °C to 42 °C. A professional UV camera, the CoroCam8, manufactured by UViRCO Technologies, has been primarily used. Based on an indicator count parameter, we have observed the temperature's impact on the corona cage's transmission line conductor. The indicator count is a numerical value generated by the ultraviolet (UV) sensor in the CoroCam8 to quantify the intensity of corona discharge. It measures UV photon hits from corona activity on energised equipment surfaces, particularly those above a threshold of around 240–280 nm UV wavelength, corresponding to ionisation events in air. Currently, no globally fixed IEC or IEEE standard defines a universal "acceptable" indicator count threshold for the CoroCam8. Several organisations (e.g., CIGRÉ, EPRI) are working toward setting guidelines for corona visual inspection and quantification.

Upon completion of the experiment setup, the voltage was elevated to +175 kV (positive), concurrently with a gradual increase in room temperature to 42 °C. The CoroCam8 is positioned 150 cm from the corona source. Figures 10–15 illustrate the spatial distribution at a constant voltage across varying temperatures. The top left corner indicates the energy loss during the experiment, known as the indicator count, along with the gain from the CoroCam8. A greater numerical value corresponds to increased losses.

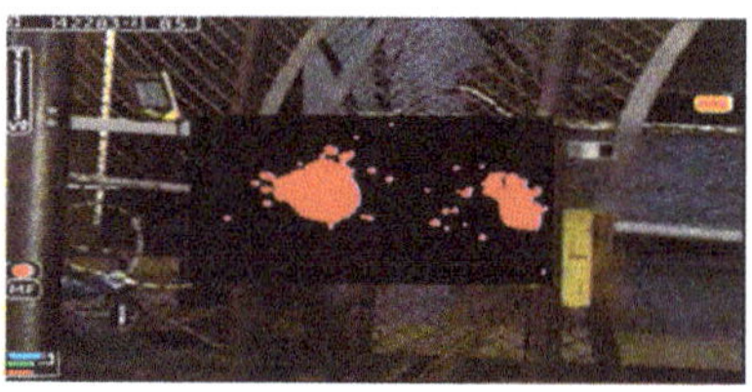

Figure 10. Positive voltage +175 kV at 25 °C.

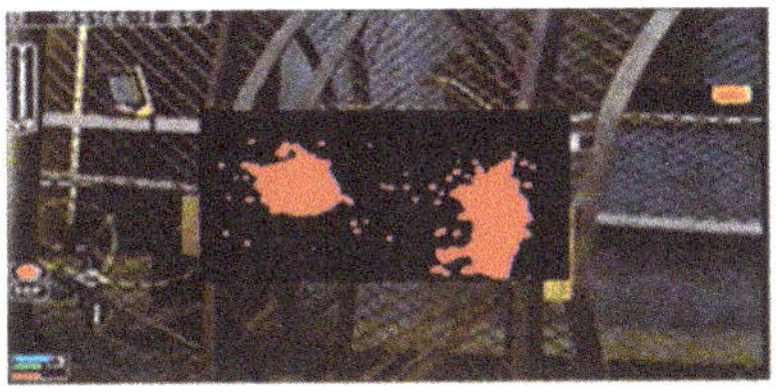

Figure 11. Positive voltage +175 kV at 30 °C.

Figure 12. Positive voltage +175 kV at 35 °C.

Figure 13. Positive voltage +175 kV at 40 °C.

Figure 14. Positive voltage +175 kV at 42 °C.

Figure 15. Breakdown at +175 kV after 42 °C.

2.5.2. Effect of Temperature on Corona-Generated Space Charge for a Negative Voltage

The supply voltage has been maintained at −175 kV. At the same time, the temperature has been slowly increased from an ambient temperature of 25 °C to 42 °C, without changing the setup's settings—the camera gains and the distance of the corocam8 from the source in the cage. The sole variable parameter, temperature, has been adjusted. Figures 16–19 depict the spatial distribution across a temperature range of 25 °C to 42 °C.

Figure 16. Negative voltage −175 kV at 25 °C.

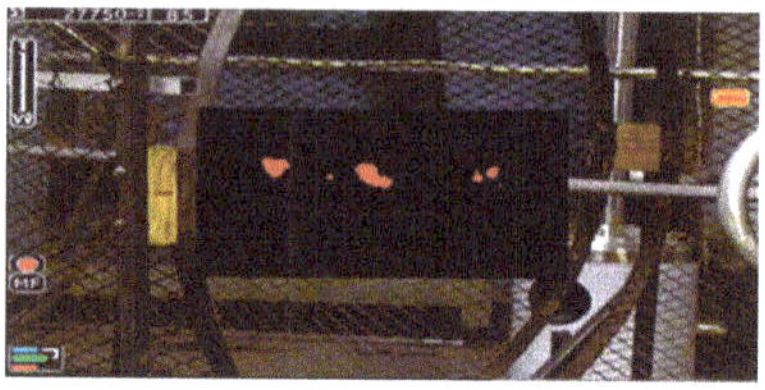

Figure 17. Negative voltage −175 kV at 30 °C.

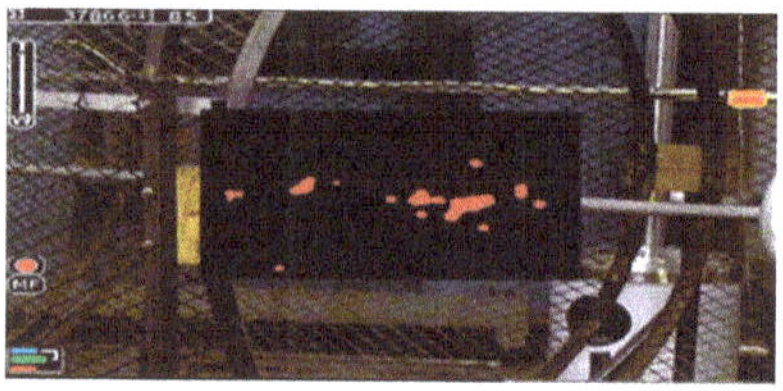

Figure 18. Negative voltage −175 kV at 35 °C.

Figure 19. Negative voltage −175 kV at 42 °C.

2.6. Effect of Temperature on Corona Loss for a Positive and Negative Voltage

Table 2 below summarises the variation in temperature and the indicator count recorded from the corocam8. It is important to note that the unit reflects the amount of energy (corona) loss during the experiment at a specific supply voltage.

Table 2. Summary indicator count vs. temperature.

Testing Voltage ±175 kV			
Positive Voltage (+V)		**Negative Voltage (−V)**	
Indicator Count	**Temperature**	**Indicator Count**	**Temperature**
126,784	21 °C	16,324	21 °C
142,283	25 °C	18,316	25 °C
195,166	30 °C	27,750	30 °C
201,833	35 °C	37,866	35 °C
218,483	40 °C	41,370	40 °C
133 (breakdown)	42 °C	43,250	42 °C

Figure 20 illustrates the variation in corona activity, measured by UV-based indicator counts from the CoroCam8 camera, as a function of ambient temperature under positive and negative HVDC voltage conditions. Under positive polarity, the indicator count increases nonlinearly with temperature, peaking around 40 °C before sharply declining at 45 °C, suggesting a possible thermal ion recombination or space charge saturation effect (breakdown). In contrast, the negative voltage condition exhibits a linear and moderate increase, reflecting a more stable space charge behaviour [32,33]. These trends highlight the polarity-dependent nature of corona discharge and suggest that atmospheric temperature plays a significant role in modulating ionisation intensity and spatial charge accumulation in HVDC environments.

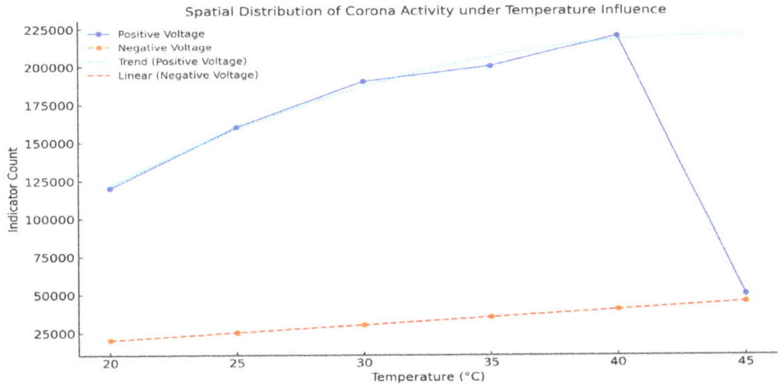

Figure 20. DC corona loss under the influence of temperature.

3. Simulation Studies

Following the experimental study, the corona-generated space charge, directly associated with the distribution of the electrical field, was derived utilising a three-dimensional model of the electrode system. The 3D model features a meshing structure including a corona cage, conductors, and rings. All elements were modelled using their precise dimensions. Figure 21 illustrates a sample model of TERN configuration. COMSOL Multiphysics 5.5 software, which is based on the finite element method, is utilised for stationary applications. Meshing is the process of dividing a complex geometry or domain into numerous, more minor, simpler pieces (finite elements) that the software can use to solve the governing equations numerically. In COMSOL Multiphysics, creating a mesh is required after building the geometry. Before running the solver, it discretises the continuous model so that COMSOL can perform a finite element analysis. The purpose of meshing is to enable the finite element method to approximate the solution on each tiny element, as the solver cannot directly apply the equations to an arbitrary, complex shape in its continuous form. A good mesh will accurately represent the geometry and provide the necessary resolution for the simulated physics. Mesh is foundational to the simulation, as it significantly influences the accuracy of the results, the convergence of the solution, and the computation time. In short, meshing bridges the gap between the CAD geometry and the numerical solver, enabling COMSOL to compute an approximate solution to physics problems in real-world geometries.

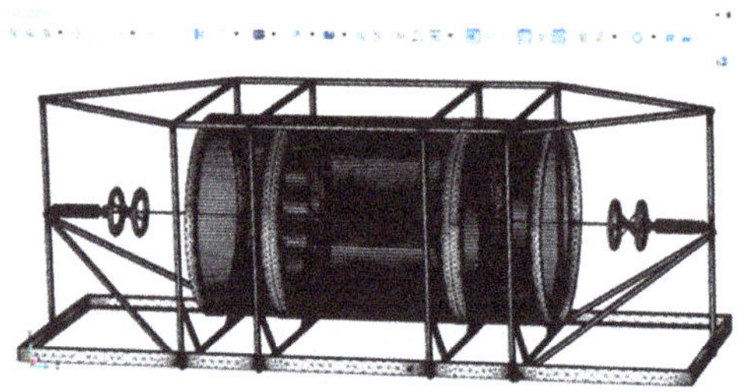

Figure 21. Meshed 3D model of the corona cage and the TERN conductor configuration.

3.1. Simulation of the Effect of Temperature on Corona-Generated Space Charge for a +V Voltage

3.1.1. Mesh Quality Evaluation

Mesh quality was assessed using COMSOL's built-in quality metrics, including aspect ratio, skewness, and element growth rate. Particular attention was paid to regions with high electric field gradients, such as near the conductor surface and ground plane, where finer meshing was applied using boundary layer refinement. The minimum element quality was maintained above 0.65 to prevent numerical instability.

3.1.2. Mesh Parameters Used

A physics-controlled mesh with user-defined element size control was applied. The conductor–air interface was discretised with a fine mesh (maximum element size: 0.005 m), while coarser elements (up to 0.05 m) were used in the outer air domain. The total mesh consisted of approximately 350,000 tetrahedral elements.

3.1.3. Mesh Convergence Test

Mesh convergence testing was performed by refining the mesh incrementally and comparing the resulting electric field magnitude at key points (e.g., conductor midpoint, corona ring tip). Convergence was confirmed when further refinement (<5% change in element size) resulted in less than 1% variation in the electric field values. The selected mesh density was, thus, validated for both accuracy and computational efficiency.

The simulation analysed temperature variation from 25 °C to 42 °C (hot atmosphere) impacts the corona discharge and electric field distribution around an HVDC transmission conductor carrying ±175 kV. The entire simulation procedure is summarised in the step-by-step setup with values, as follows:

Step 1: Start a New COMSOL Project

- Choose Space Dimension: 3D. Next, select Physics Interfaces.

Step 2: Add Physics:

- AC/DC Module > Electrostatics (for electric field distribution).
- Plasma Module > Corona Discharge (to simulate corona discharge and space charge).
- Heat Transfer Module > Heat Transfer in Fluids (to model temperature variation).

Step 3: Define the Geometry

- HVDC Transmission Line Conductor Geometry:
    - Create a cylinder to represent the HVDC conductor.
    - Radius = 2 cm (0.02 m).
    - Length = 2 m (arbitrary, choose the length to simulate a section of the line).
- Air Domain:

- ■ Create a larger cylinder representing the air surrounding the HVDC conductor line.
- Radius = 0.75 m (to capture the influence of the electric field over a reasonable distance), the size of the cage.
- Height = 2 m (same as the conductor length).
  - ■ Ground Plane:
    - Add a flat plane below the conductor (at 75 cm) to simulate the ground.

Step 4: Define Materials

- Conductor: Assign aluminium to the HVDC line (from the Material Library).
- Relative permittivity (dielectric constant) = 1 (for metals).
- Conductivity = 5.8 × 1075.8\times 10^75.8 × 107 S/m.
  - ■ Air Domain: Assign air to the surrounding volume.
- Relative permittivity = 1 (for air).
- Electrical conductivity: Use default air properties.

Step 5: Setup Electrostatics (AC/DC Module)

- Voltage Potential:
  - ■ Select the conductor surface and set the electric potential to ±175 kV.
  - ■ Boundary condition: 200 kV DC (±200,000 V).
- Ground Potential:
- Set the ground plane (bottom of the air domain) to 0 V (ground potential).
  - ■ Boundary Conditions:
- Leave the rest of the air domain as open boundary conditions, allowing the field to decay naturally in space.

Step 6: Setup Corona Discharge (Plasma Module)

- Activate Corona Discharge:
- Apply the Corona Discharge interface to the air domain.
  - ■ Corona Discharge Settings:
- Corona inception voltage: Set the threshold voltage for corona discharge to 200 kV (typical value for air breakdown).
  - ■ Positive and Negative Ions:
- For the ion species, define:
  - ■ Positive ion mobility = 1.5 × 10 − 41.5 \times 10{−4}1.5 × 10^{-4} m^2/Vs (for typical ions in air).
  - ■ Negative ion mobility = 2.0 × 10 − 42.0\times 10{−4}2.0 × 10^{-4} m^2/Vs.
- Space Charge Accumulation:
  - ■ Add positive and negative ion densities to simulate charge build-up.
    - Initial space charge density: 0 C/m^3 (assume no initial charge).

Step 7: Setup Heat Transfer (Heat Transfer Module)

- Heat Transfer in Air:
  - ■ Apply the Heat Transfer in the Fluids interface to the air domain.
- Initial Temperature:
  - ■ Set the initial air temperature to 25 °C, standard or ambient temperature
- Variable Temperature:

  - In the parametric sweep (Step 10), we varied the temperature between 25 °C and 42 °C.
- Air Properties:
  - Air properties, such as thermal conductivity, density, and specific heat, should be taken from COMSOL's air material properties. These are already temperature-dependent in COMSOL.

Step 8: Multiphysics Coupling

- Electrostatics and Corona Discharge:
  - Couple the Electrostatics with the Corona Discharge module to ensure that the corona-generated space charges affect the electric field distribution.
- Heat Transfer and Corona Discharge:
  - Couple Heat Transfer and Corona Discharge, so that the air's ion mobility is affected by temperature, where mobility increases.

Step 9: Mesh Generation

- Mesh Settings:
  - Use a fine mesh near the conductor to accurately capture the electric field and corona-generated space charge effects.
  - A coarse mesh can be used for the outer parts of the air domain to save computation time.

Step 10: Study Setup

- Study Type: Select Stationary Study (for a steady-state solution).
- Parametric Sweep:
  - Define a parametric sweep for 25 °C and 42 °C atmospheric temperatures.
    - Parametric values: 25 °C, 30 °C, 35 °C, and 42 °C.
- Compute: Run the simulation.

After a proper setup and adjustment of the cage geometry and all other settings, the simulation was run for every specific temperature, with the voltages maintained at ±175 kV. The display of the corona generated around the conductor is shown in Figures 22–29 below.

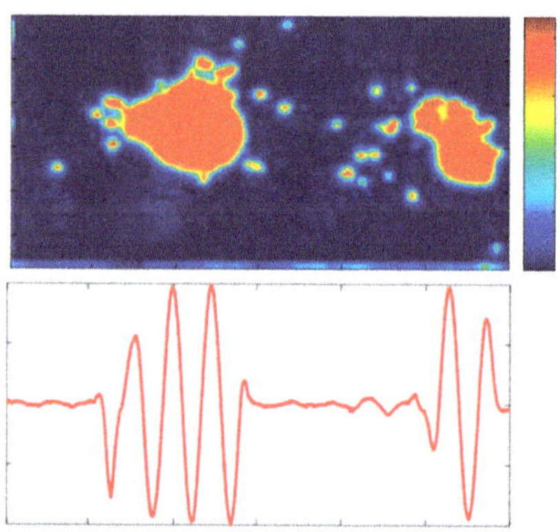

Figure 22. Positive voltage +175 kV at 25 °C.

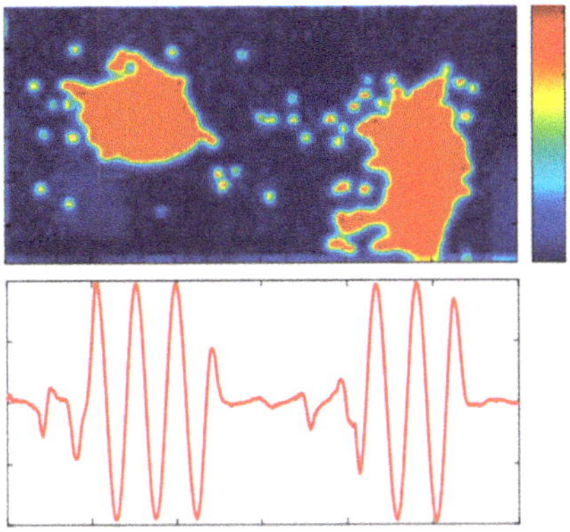

Figure 23. Positive voltage +175 kV at 30 °C.

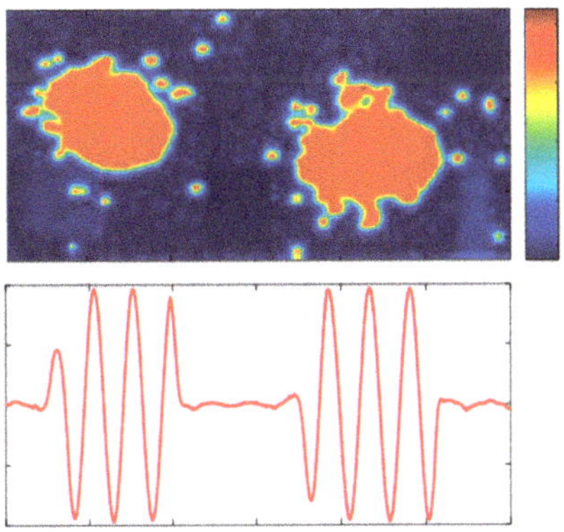

Figure 24. Positive voltage +175 kV at 35 °C.

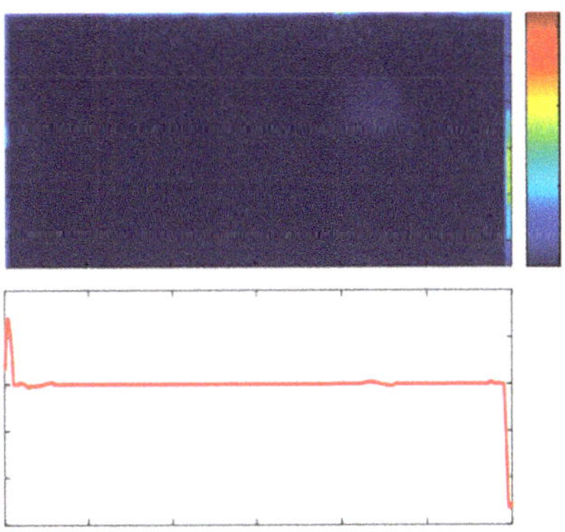

Figure 25. Breakdown stage +175 kV after 42 °C.

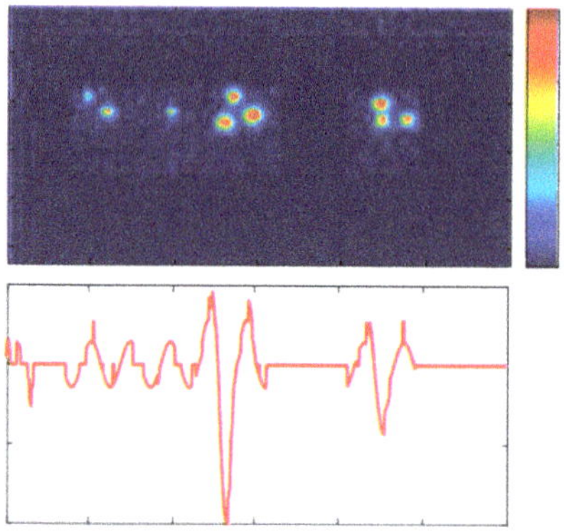

Figure 26. Negative voltage −175 kV at 25 °C.

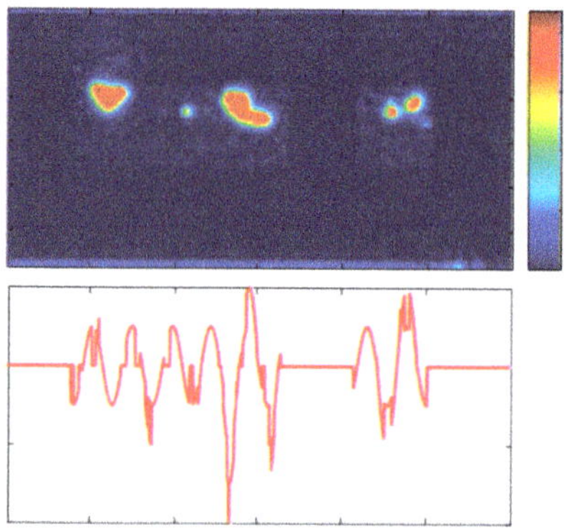

Figure 27. Negative voltage −175 kV at 30 °C.

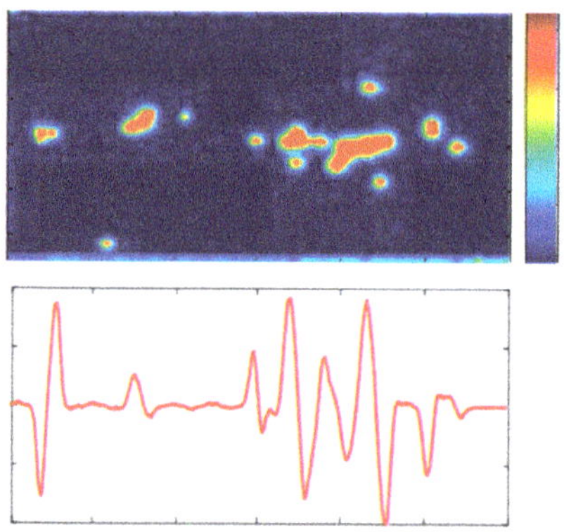

Figure 28. Negative voltage −175 kV at 35 °C.

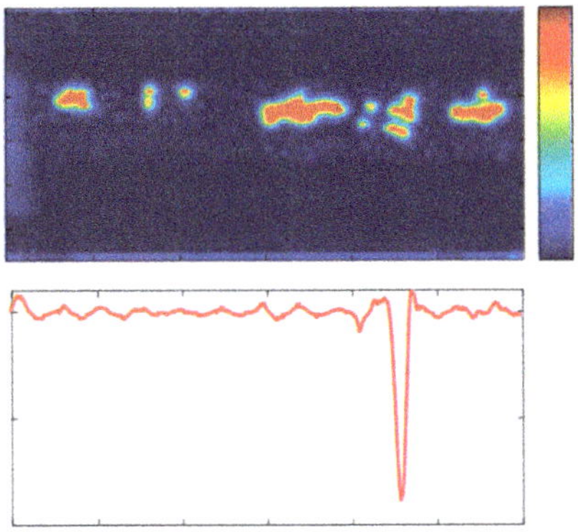

Figure 29. Negative voltage −175 kV at 42 °C.

3.2. Simulation of the Effect of Temperature on Corona-Generated Space Charge for a −V Voltage

For the negative voltage simulation case, after changing the polarity of the voltage to −175 kV and increasing the temperature from ambient 25 °C to 42 °C, Figures 26–29 illustrate the effect of temperature on corona accumulation.

3.3. Validation Strategy: Simulation vs. Experimental Results

The electric field and space charge distributions were validated through comparison with experimental data collected in the indoor corona cage setup, as follows:

3.3.1. Electric Field Validation

Electric field probes were simulated at the same spatial points where field sensors or CoroCam8 camera readings were focused.

The electric field magnitude at the conductor surface and cage midplane was compared to inferred field values from corona onset voltages.

3.3.2. Space Charge Validation

Indicators count values from the CoroCam8 camera were used to proxy ionisation intensity. Simulated space charge densities (in C/m^3) were mapped spatially and correlated with camera-detected hotspots. Regions of high simulated space charge concentration overlapped with areas of elevated UV emission in the CoroCam8 recordings, confirming spatial consistency.

3.3.3. Quantitative Alignment

Voltage thresholds at which corona first appeared in the experiment (e.g., +21 kV and 16 kV) were matched in simulation by adjusting the temperature, air density, and electric field strength until corona onset (i.e., ionization) occurred in the model. Good agreement (within 5–10%) was achieved in electric field values and indicator count trends, particularly at temperatures in the range of 25–39 °C.

3.4. Simulation Limitations (Thermal Modelling)

Below are the limitations in the FEM simulations related to thermal effects:

- The FEM simulations assumed uniform temperature fields, which may not reflect localised thermal gradients within the corona cage.
- The thermal coupling between the ionized air and surrounding surfaces was simplified, omitting transient or convective effects.

- Humidity was excluded as a parameter, though it is known to influence breakdown strength and charge mobility.
- The lack of direct space charge measurement tools means validation depends on indirect indicators (e.g., UV count, electric field).

4. Conclusions

This article analyses DC corona characteristics for a specific conductor, called the "TERN" conductor, utilising both experimental and simulation methods. An HVDC system was initially modelled using a corona cage with a scaled conductor. A cylindrical indoor corona cage measuring 270 cm in length and 150 cm in diameter was utilised for this purpose. After the initial assessments conducted with the single-line configurations, measurements were obtained using a daylight UV corona camera. During the second phase of the study, simulations based on the finite element method (FEM) were conducted utilising COMSOL Multiphysics software to ascertain the electric field distributions for the specified conductor across various atmospheric temperatures.

Simulations and test results were utilised alongside empirical formulas to assess the characteristics of the remaining corona-generated space charge. The test results are subsequently compared and analysed. The experimental results align with the findings from simulation studies. The corona test results were obtained by measuring the inception voltage using the conventional method of apparent charge measurement on a corona cage with three sections under both positive and negative voltages. The results show that the inception voltage and magnitude of the corona-generated space charge decrease very little with the increase in temperature under a DC voltage supply, and the negative inception voltage is lower than the positive inception voltage. The corona-generated space charge is related to the electric field; the loss is more considerable under the influence of temperature with a positive voltage (+V) than a negative voltage (−V).

The study confirms that corona-generated space charges significantly impact the electric field distribution in an indoor HVDC corona cage, with atmospheric temperature playing a key role as an influencing factor. Elevated temperatures enhance ion mobility and reduce air density, increasing space charge accumulation and localised field distortion. These results highlight the necessity of considering thermal effects when analysing field behaviour in HVDC environments, especially for accurate testing and insulation evaluation.

Author Contributions: Methodology, J.S.D.; software, J.S.D.; writing—review and editing, T.S.; visualization, J.S.D.; supervision, H.M.L.; project administration, H.M.L. All authors have read and agreed to the published version of the manuscript.

Funding: This research received no external funding.

Data Availability Statement: Data are contained within the article.

Conflicts of Interest: The authors declare no conflict of interest.

References

1. Dallaire, R.D.; Maruvada, P.S.; Rivest, N. HVDC Monopolar and Bipolar Cage Studies on the Corona Performance of Conductor Bundles. *IEEE Trans. Power Appar. Syst.* **1984**, *PAS-103*, 84–91. [CrossRef]
2. Wang, F.; Zheng, H.; Shao, T. Study of the Corona Characteristics and Space Charge Distribution of HVDC Transmission Lines. *IEEE Access* **2020**, *8*, 131–137.
3. Fang, M.; Wang, L.; Zhan, Y. Temperature-Dependent Characteristics of Space Charge Accumulation in Gas-Insulated HVDC Systems. *IEEE Trans. Power Deliv.* **2022**, *37*, 2712–2721.
4. Liu, C.; Zhang, Y.; Zhang, H.; Shi, Q. Effect of Temperature on Corona Discharge and Ion Flow Characteristics in HVDC Substations. *IEEE Trans. Power Deliv.* **2021**, *36*, 1692–1700.

5. Chen, Y.; Zhou, Y.; Jiang, Z. Modeling and Analysis of Corona-Induced Space Charge in Indoor HVDC Test Chambers. *IEEE Trans. Dielectr. Electr. Insul.* **2022**, *29*, 2201–2209.
6. Yi, Y.; Zhang, C.; Wang, L. Positive DC corona inception on dielectric coated stranded conductors in air. *IET Sci. Meas. Technol.* **2016**, *10*, 557–563. [CrossRef]
7. Bas-Calopa, P.; Riba, J.R.; Moreno-Eguilaz, M. Measurement of corona discharges under variable geometry, frequency, and pressure environment. *Sensors* **2022**, *22*, 1856. [CrossRef]
8. Gharaaghaji, S.; Fadil, H.; Ilhan, S.; Ozdemir, A.; Ismailoglu, H.; Cortes, F.E. Corona Inception and Breakdown Voltages of Rod–Plane Electrode for Severe Ambient Conditions. In Proceedings of the 2019 IEEE PowerTech Conference, Milan, Italy, 23–27 June 2019; IEEE: New York, NY, USA, 2019.
9. Xiao, F.; Zhang, B.; Deng, Y. Experimental Study on Impulse Corona Characteristics of HVDC Conductors With Space Charges. *IEEE Trans. Power Deliv.* **2021**, *36*, 1705–1713. [CrossRef]
10. Zhao, W.; Zhang, J.; Wang, L. Simulation of Space Charge Effect on Electric Field Distribution in DC GIS Under Temperature Variation. *IEEE Trans. Plasma Sci.* **2020**, *48*, 3260–3266.
11. Wang, H.; Liu, J.; Xie, R. Numerical Analysis of Space Charge Effects on HVDC Electric Field Distribution Considering Environmental Parameters. *IEEE Trans. Dielectr. Electr. Insul.* **2021**, *28*, 1753–1761.
12. Djeumen, J.S.; Walker, J.J.; West, N.J. Measurement of space charge distribution in a corona cage under the influence of different atmospheric conditions. In Proceedings of the 2018 IEEE International Conference on High Voltage Engineering and Application (ICHVE 2018), Athens, Greece, 10–13 September 2018; p. 107.
13. Xu, Z.; Li, B.; Wang, D. Temperature Influence on Space Charge Distribution and Breakdown Strength in Epoxy Composites for HVDC Applications. *IEEE Trans. Dielectr. Electr. Insul.* **2020**, *27*, 1050–1057.
14. Djeumen, J.S.; Walker, J.J. Environmental influence on corona inception with HVDC application. In Proceedings of the International Symposium on High Voltage Engineering, (ISH 2017), Buenos Aires, Argentina, 28 August–1 September 2017; pp. 1–6.
15. Yang, K.; Liu, Q.; Chen, L. Experimental Study on Electric Field Distortion Caused by Corona Discharge in HVDC Chambers. *IEEE Access* **2021**, *9*, 78823–78831.
16. Megala, V.; Karpagam, R.; Sathishkumar, G.K.; Gopinath, B.; Marish Kumar, P.; Deva Brindha, M. Investigation on corona performance of conductors using fabricated indoor corona cage. *Mater. Today Proc.* **2021**, *44*, 3652–3656. [CrossRef]
17. Nayak, M.; Sekhar, K.C. Computation of corona-generated ionic current environment of unipolar UHVDC transmission lines using gas bimolecular collision theory. *Electr. Power Syst. Res.* **2024**, *230*, 110204. [CrossRef]
18. Liu, H.; Liao, R.; Zhao, X.; Lin, Y. Variation of surface electric field intensity determined by space charge density at different temperatures. *IEEE Trans. Dielectr. Electr. Insul.* **2019**, *26*, 1660–1668. [CrossRef]
19. Bian, X.B.; Liming, W.; Zhicheng, G.; Jing, C.; Yingjian, Y.; Xiong, W. Experimental investigation on altitude correction factor of positive dc corona inception voltages of transmission lines based on the mobile corona cage. In Proceedings of the International Conference High Voltage Engineering (ICHVE), New Orleans, LA, USA, 11–14 October 2010; pp. 548–551.
20. Font, A.; Solak, Y.T.; Ilhan, S.; Ozdemir, A.; Ismailoglu, H.; Cortes, F.E. DC corona characteristics derived from corona cage tests and simulations. *IEEE Trans. Dielectr. Electr. Insul.* **2019**, *26*, 1867–1875. [CrossRef]
21. Zhang, B.; Wang, W.; He, J. Theoretical Study on Radio Interference of HVDC Transmission Line Based on Cage Tests. *IEEE Trans. Power Deliv.* **2017**, *32*, 1891–1898. [CrossRef]
22. Xiao, F.; Zhang, B. Transient Overvoltage on HVDC Overhead Transmission Lines With Background DC Space Charges and Impulse Corona. *IEEE Trans. Power Deliv.* **2021**, *36*, 2921–2928. [CrossRef]
23. Nayak, M.R.; Sekhar, K.C. A Novel Computational Technique to Analyse the Corona Generated Ionized Field Environment of EHV/UHV DC Transmission Lines. *Int. J. Emerg. Electr. Power Syst.* **2023**, *25*, 667–674.
24. Nayak, M.R.; Mujeer, S.A. New Computational Method for Study of Ionic Current Environment of HVDC Transmission Lines. In Proceedings of the 2020 IEEE International Conference on Advances and Developments in Electrical and Electronics Engineering (ICADEE), Coimbatore, India, 10–11 December 2020.
25. Maruvada, P.S.; Dallaire, R.D.; Héroux, P.; Rivest, N. Corona Studies for Biploar HVDC Transmission at Voltages Between ±600 kV AND ±1200 kV PART 2: Special Biploar Line, Bipolar Cage and Bus Studies. *IEEE Trans. Power Appar. Syst.* **1981**, *100*, 1462–1471. [CrossRef]
26. Smith, J.; Lee, A. High Voltage Direct Current (HVDC) Transmission. *J. Energy Nat. Resour. Res.* **2023**, *5*, 45–56.
27. Kyere, I.K.; Nyamupangedengu, C.; Swanson, A.G. A Comparative Study of Time-Evolution Characteristics of Single and Double Cavity Partial Discharges. *Energies* **2024**, *17*, 1905. [CrossRef]
28. Doe, J.; Zhang, Y. Advancements and Comparative Analysis of High-Voltage Direct Current Transmission Technologies. *Int. J. Electr. Power Energy Syst.* **2023**, *120*, 100–110.
29. Mahamuda, M.; Yoshida, K.; Suzuki, Y.; Murakami, K. Influence of Conductor Temperature on the V–I Characteristic of Corona Discharge in a Coaxial Arrangement: Experiments and Simulation. *Energies* **2025**, *18*, 1303.

30. Sibilant, G. A Study of HVDC Conductor Corona in a Purpose-Built Corona Cage. Master's Thesis, University of Durban-Westville, Westville, South Africa, 2003.
31. Swanson, A.; Ijumba, N.; Stephen, R. Experimental Investigation of the Effect of Conductor Temperature on Corona Performance in Overhead Transmission Lines. *IEEE Trans. Power Deliv.* **2024**, *38*, 952–960.
32. He, K.; Gu, J.; Huang, S.; Bian, K.; Ju, Y.; Chen, W.; Lu, J. The Corona Space Charge Distribution of Changji-Guquan ±1100kV UHVDC Transmission Line within the Thundercloud Electric Field. In Proceedings of the 2020 IEEE International Conference on High Voltage Engineering and Application, Beijing, China, 6–10 September 2020.
33. Pfeiffer, M.; Hedtke, S.; Franck, C. Corona Current Coupling in Bipolar HVDC and Hybrid HVAC/HVDC Overhead Lines. *IEEE Trans. Power Deliv.* **2018**, *33*, 393–402. [CrossRef]

energies

MDPI

Article

A Simulation and a Computational Study on the Reliability Verification of Epoxy Resin Paper-Impregnated Bushings in Power Transformers

Daijun Liu [1], Xiaobang Tong [2], Libao Liu [2], Xiaoying Dong [1], Tianming Yan [1], Wenkai Tang [2], Liming Wang [3], Bin Cao [3,*] and Zimin Luo [3,*]

[1] China Yangtze Power Co., Ltd., Beijing 100032, China
[2] Nanjing Electric High Voltage Bushing Co., Ltd., Nanjing 210046, China
[3] Shenzhen International Graduate School, Tsinghua University, Shenzhen 518055, China
* Correspondence: cao.bin@sz.tsinghua.edu.cn (B.C.); luozm23@mails.tsinghua.edu.cn (Z.L.)

Abstract: Epoxy resin paper-impregnated bushings, as critical insulating components in power transformers, are subjected to complex electric fields, thermal fields, and mechanical stresses over extended periods. Their performance stability is directly linked to the safe operation of transformers. Given the significant costs associated with their production, reliability verification is a crucial aspect of their design and manufacturing process. This study employs the finite element simulation technology to systematically investigate the electric field distribution characteristics, thermal field distribution characteristics, and seismic performance reliability verification methods of epoxy resin paper-impregnated bushings. The simulation and calculation results indicate that for bushings with rated voltages of 40.5 kV, 72.5 kV, and 126 kV, the maximum radial electric field strengths are 1.38 kV/mm, 2.74 kV/mm, and 3.0 kV/mm, respectively, with axial electric field strengths all below allowable values. The insulation margin meets the 1.5 standard requirements. Under short-circuit conditions, the thermal stability analysis of the bushings reveals that the final conductor temperatures are all below 180 °C, indicating sufficient safety margins. All three types of bushings comply with the design requirements for an 8-degree earthquake intensity and are capable of effectively withstanding seismic loads. This research provides a theoretical foundation for the development and application of epoxy resin paper-impregnated bushings, offering a significant engineering application value in enhancing the safety and stability of transformers and power systems.

Keywords: epoxy resin; paper-impregnated bushings; electric field strength; seismic performance; thermal field analysis

1. Introduction

As core insulating components in power transformers, the bushings primarily serve insulation and support functions and play a significant role in power systems [1–3]. With the continuous increase in voltage levels in power systems and the development of power equipment toward high capacity and high reliability, the performance stability of epoxy resin paper-impregnated bushings directly impacts the safe operation of transformers and the entire power system [4,5]. However, the research and production costs of epoxy resin paper-impregnated bushings are high, and during long-term operation, they are subjected to the combined effects of electric field, thermal field, and mechanical stresses. Therefore, conducting scientific reliability verification is of great significance [6,7].

In recent years, epoxy resin paper-impregnated bushings have gained prominence as replacements for traditional oil-paper bushings in power systems, owing to their superior electrical and mechanical properties [7,8]. However, due to the complex physical characteristics of epoxy resin materials, these bushings may encounter challenges such as uneven electric field distribution, thermal accumulation, and mechanical damage under extreme operating conditions like earthquakes [6,7,9,10]. The non-uniform electric field distribution is the core risk factor that causes various physical and chemical deterioration processes and significantly reduces the operational reliability of epoxy resin-impregnated gaskets during actual operation. The non-uniform electric field will cause the electric field in certain areas to be excessively high and prone to inducing spatial charge accumulation, local discharge, and aging of electrical tree branches [1]. The heat generated by local discharge and active particles will further damage the epoxy material, and through multi-stress coupling effects such as thermal–electrical–mechanical–chemical effects, accelerate the insulation aging process [2,3]. The strong shock and vibration brought by earthquakes will cause the gasket to bear mechanical loads far exceeding those during normal operation. In high-humidity and hot regions, the epoxy resin and corrugated paper will absorb moisture, which increases their conductivity and dielectric loss, making them more prone to loss of heat and breakdown problems.

Consequently, the scientific verification of their reliability, the optimization of design parameters, and the enhancement of their overall performance have become focal points in the current research landscape of power equipment development. Existing studies have demonstrated that the finite element simulation technology can effectively analyze the electric field distribution characteristics of bushings and optimize their insulation structure design [11,12]. Based on the actual geometric shape, material properties, and operating conditions of the power transformer, detailed models of the gasket and its surrounding components are established. By setting accurate mathematical models and boundary conditions, finite element simulation can accurately predict the temperature and mechanical stress distribution of the equipment during actual operation [13]. Comparing the actual working environment of the high-voltage direct current bushing with the temperature and mechanical stress tolerance capabilities of the material, the thermal stability and mechanical stability of the bushing can be predicted.

Currently, researchers worldwide have achieved certain accomplishments in the performance study of epoxy resin paper-impregnated bushings. Xu and others established an electric field simulation model for epoxy resin paper-impregnated bushings after faults, analyzing the electric field distribution under normal operation, insulation breakdown, and internal bubble conditions. Simulation results revealed that bushing explosion faults are caused by the combined effects of bubbles, moisture penetration, bushing head deformation, and short-circuit current impacts [14]. Thermal field analysis methods have also been widely applied to assess the thermal stability of bushings. Yadav and others improved the thermal resistance of bushings by modifying epoxy resin with nano hexagonal boron nitride and adding nanocellulose fibers, achieving a thermal performance improvement of approximately 4.19%. The proposed neural network model effectively predicts temperature distribution with minimal deviation (<2%) compared to experimental values, providing a reliable method for optimizing high-voltage bushing heat dissipation management [15]. Seismic performance studies have also provided theoretical support for the application of bushings in earthquake-prone regions [15,16]. However, existing research primarily focuses on the performance variation patterns of bushing equipment under different defects, while the integrated reliability verification methods for electric field, thermal field, and seismic performance designs remain incomplete. Systematic reliability verification techniques for epoxy resin paper-impregnated bushings under high voltage levels and complex operating conditions still require further exploration.

This study is based on the finite element simulation technology and theoretical calculation methods, systematically investigating the reliability verification methods for epoxy resin paper-impregnated bushings in terms of their electric field distribution characteristics, thermal field distribution characteristics, and seismic performance. The research aims to provide a theoretical foundation for the development and engineering application of high-voltage bushings. The study encompasses a comprehensive analysis of the bushings' electric field distribution, thermal field distribution, and seismic performance, as well as an in-depth exploration of their reliability verification methods. This research will offer scientific guidance for enhancing the design level and reliability of epoxy resin paper-impregnated bushings, contributing to the safety and stability of transformers and power systems, and holds significant engineering application value.

2. Simulation and Test Methods

2.1. Geometric Model

The structures of the three bushings with different voltage ratings are illustrated in Figure 1. The 40.5 kV, 72.5 kV, and 126 kV bushings all adopt an integral current-carrying structure where the winding rod serves as the current-carrying rod. The primary components include the air-side terminal plate, porcelain housing, main flange, epoxy resin-impregnated paper core, and oil-side terminal plate. The main flange of the bushing is bonded to the core, and the porcelain housing is securely fastened using a Belleville spring clamp. The space between the porcelain housing and the core is filled with insulating paste, forming the outer insulation. The oil-side terminal plate is threaded onto the conductive rod and secured with a locking nut. The 72.5 kV bushing incorporates additional components such as a metallic washer, a connector assembly, and a voltage balancing sphere, with a clamping force of approximately 16 kN. For the 126 kV bushing, additional elements include a conductive head, a voltage balancing seal cover, and a conductive rod. Current flow is achieved at the head through two spring contact fingers between the conductive tube and the conductive head.

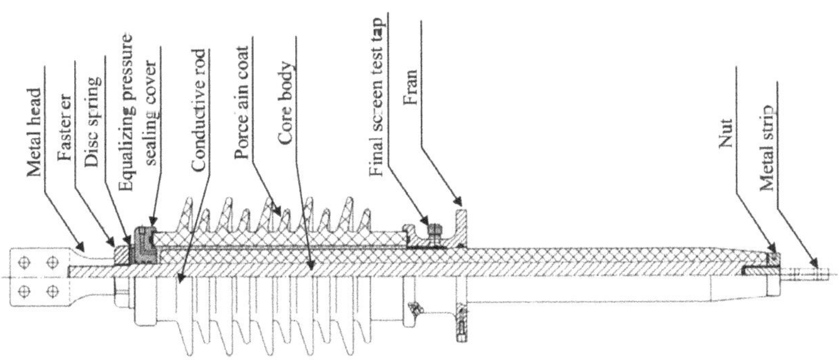

(**a**) The 40.5 kV bushing structure

Figure 1. *Cont.*

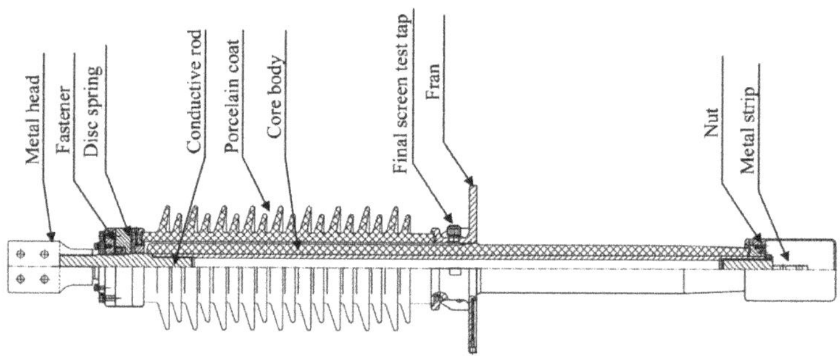

(**b**) The 72.5 kV bushing structure

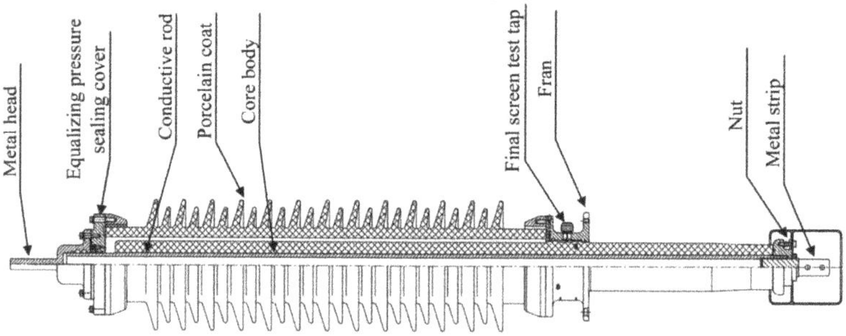

(**c**) The 126 kV bushing structure

Figure 1. A schematic diagram of bushing structures with different voltage levels.

Under the dual requirements of addressing field strength safety and process suitability, the inter-screen voltage distribution is coordinated with the partial discharge inception voltage to ensure that the bushing's insulation structure and material properties align with operational conditions. A design methodology employing uniform thickness with graded steps is implemented, where axial field strength is maintained below permissible limits by regulating umbrella-step and tail-step differentials to control step dimensions, thereby achieving sufficient insulation design margins. The 40.5 kV, 72.5 kV, and 126 kV bushings incorporate 7, 8, and 10 capacitive grading screens, respectively. The simulation model of the 126 kV bushing, as illustrated in Figure 2, serves as a representative example. The dimensions of bushing screens for three different voltage levels of bushings are shown in Table 1.

Table 1. Dimensions of bushing screens for three different voltage levels of bushings.

Voltage Level		40.5 kV	72.5 kV	126 kV
Capacitance core radius (cm)		4.6	11.6	13
Screen spacing (cm)		0.25	0.225	0.225
Length of screen (cm)	0 #	114.7	147	185
	1 #	98.7	125.7	154.5
	2 #	88.7	116.1	145.5
	3 #	80.3	109.1	136.7
	4 #	74	101.9	127.9
	5 #	67.6	94.6	119.1

Table 1. *Cont.*

Voltage Level	40.5 kV	72.5 kV	126 kV
6 #	60.9	87	110.1
7 #	53.7	79	100.8
8 #	/	70.3	91.1
9 #	/	/	83.5
10 #	/	/	75.3

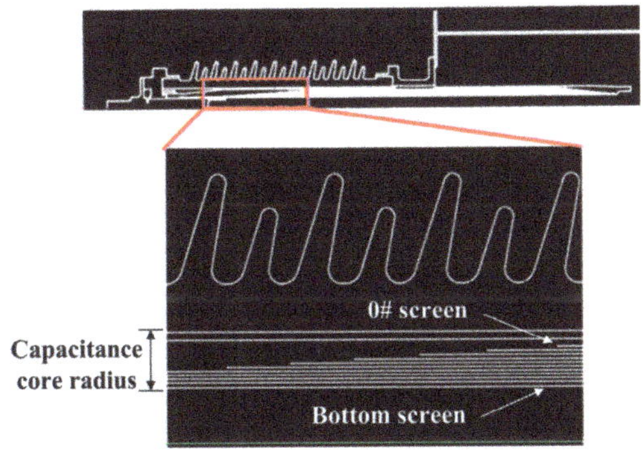

Figure 2. A structural representation of the 126 kV bushing simulation model.

2.2. *Mathematical Models*

2.2.1. A Mathematical Model for Electric Field Simulation

The bushings studied in this paper are subjected to the combined effects of AC and DC voltages, resulting in electric field distributions in the epoxy resin capacitor core that are influenced by both conductivity and permittivity [14,17]. The simulation frequency of the electric field of the bushing is 50 Hz. At this time, the electromagnetic wave length is much larger than the size of the bushing, and the displacement current is much smaller than the conducted current, meeting the quasi-static condition. At this point, the time-varying effect of the magnetic field can be ignored, and the electric field can be solved independently. The fundamental equations governing electric field distribution can be derived and solved using Maxwell's equations.

$$\nabla \cdot \boldsymbol{D} - \rho_e \quad (1)$$

$$\nabla \times \boldsymbol{E} \approx \boldsymbol{O} \quad (2)$$

$$\nabla \times \boldsymbol{H} = J + \frac{\partial D}{\partial t} \quad (3)$$

$$\boldsymbol{D} = \varepsilon \boldsymbol{E} \quad (4)$$

$$J = \gamma \boldsymbol{E} \quad (5)$$

$$\boldsymbol{E} = -\nabla \varphi \quad (6)$$

where D is the electric displacement vector, $\mathrm{C/m^2}$; E is the electric field intensity, V/m; ρ_e is the charge density, $\mathrm{C/m^3}$; H is magnetic field strength, $\mathrm{A/m}$; J is current density, $\mathrm{A/m^2}$; ε is permittivity, $\mathrm{F/m}$; γ represents the electrical conductivity of the material, indicating the ability of the material to conduct electric current, $\mathrm{S/m}$; φ is electric potential, V.

By combining Equations (1), (2), and (4), we can obtain:

$$\nabla \cdot \varepsilon(-\nabla \varphi) = \rho_e \quad (7)$$

$$\nabla^2 \varphi = -\frac{\rho_e}{\varepsilon} \tag{8}$$

Epoxy resin-impregnated paper can be regarded as an ideal medium without free charge at power frequency. Therefore, in Equation (8), it is directly assumed that $\rho_e = 0$. When free charge effects are neglected in the computational domain, the governing equation is reduced to:

$$\nabla^2 \varphi = 0 \tag{9}$$

Under operational conditions, bushings are subjected to combined AC/DC voltages, leading to time-varying electric fields. To calculate the electric field under such conditions, the displacement-induced electric field and steady-state conduction field must be superimposed, yielding the following governing equation:

$$\nabla \cdot \boldsymbol{J} = -\frac{\partial \rho_e}{\partial t} \tag{10}$$

Then, we can obtain:

$$\nabla \cdot \left(r\boldsymbol{E} + \varepsilon \frac{\partial \boldsymbol{E}}{\partial t} \right) = 0 \tag{11}$$

Substituting Equation (6) into Equation (11) yields:

$$\left(\gamma + \varepsilon \frac{\partial}{\partial t} \right) \nabla^2 \varphi = 0 \tag{12}$$

Equation (12) accounts for the effects of conduction current and displacement current on the electric field, enabling the determination of the electric field distribution. This study incorporates temperature dependence, where the conductivity γ is defined as a function of T:

$$\gamma = 2.21 \times 10^{-7} e^{\left(-\frac{4.826}{T}\right)} \tag{13}$$

where γ is the electrical conductivity of the material used in the simulation, indicating the material's ability to conduct current, S/m; T is temperature, °C.

2.2.2. A Mathematical Model for Seismic Performance

According to IEC TS 61463 [18], the seismic capability of bushings is verified by primarily assessing whether the failure torque of the fiberglass tube at its critical section can meet (exceed) the total bending moment occurring during an earthquake (which includes bending moments caused by seismic loads, wind loads, and terminal loads). The seismic calculation diagrams for bushings of different voltage levels are illustrated in Figure 3.

The horizontal forces, vertical forces, pressure, and bending moments induced by seismic loads can be calculated using Equations (14)–(17).

$$F_h = m_p \times K \times a_{gh} \times R \times S_c \tag{14}$$

$$F_v = m_p \times K \times a_{gv} \times R \times S_c + m_p g \tag{15}$$

$$F_a = -F_h \times \sin\alpha + F_v \times \cos\alpha \tag{16}$$

$$F_b = -F_h \times \cos\alpha + F_v \times \sin\alpha \tag{17}$$

where a_{gh} is the horizontal ground acceleration, m/s^2; a_{gv} is the vertical ground acceleration, m/s^2; the horizontal acceleration of the transformer tank is $K \times a_{gh}$, and the vertical acceleration of the transformer tank is $K \times a_{gv}$, where K is an amplification factor, taken as 2; m_p is the mass of the bushing, kg; α is the installation angle in the vertical plane; R

is the response coefficient, taken as 2.5; S_c is the coefficient considering multi-frequency excitation and multi-mode response, taken as 1.5.

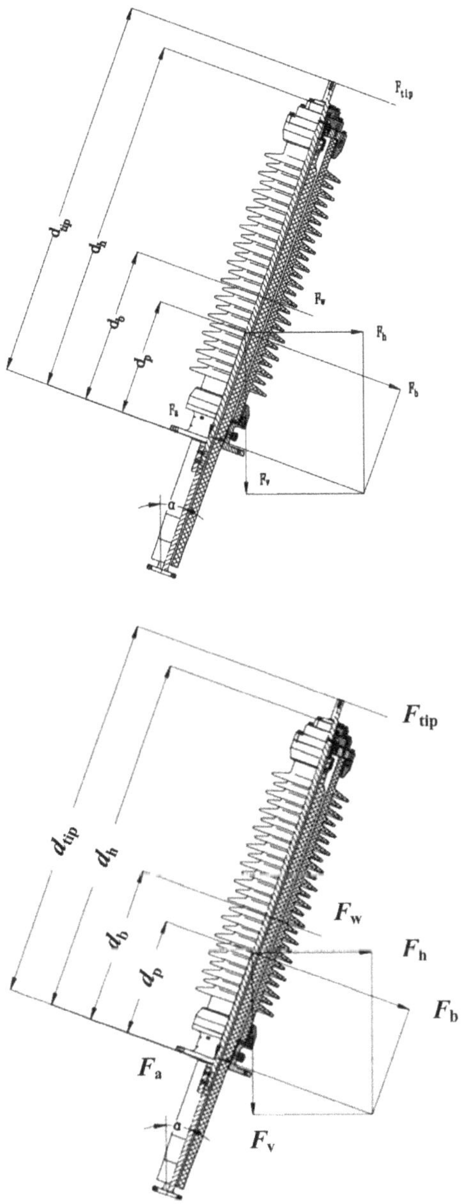

Figure 3. A structure diagram of bushings for seismic calculation.

Then, the seismic- and gravitation-induced bending moments can be calculated using Equation (18), where F_b is the bending force, N; d_p is the distance between the centroid of the bushing and the critical cross-section, mm.

$$M_{bs} = F_b \times d_p \tag{18}$$

The wind load calculation is given by Equation (19), and the bending load caused by wind load can be calculated using Equation (20).

$$F_w = P \times (D_e + D_i) \times d_h \div 2 \tag{19}$$

$$M_{bw} = F_w \times d_b \tag{20}$$

Here, P represents wind pressure, Pa; D_e is the maximum outer diameter of the umbrella skirt, mm; D_i is the outer diameter of the umbrella core, mm; d_h is the distance from the critical cross-section to the top of the bushing, mm; F_w is wind force, N; and d_b is the centroid at the gas end, mm.

The equation for the bending moment caused by terminal load is provided by Equation (21), where d_{tip} denotes the distance from the critical cross-section to the terminal, mm; F_{op} is the cantilever working load, N.

$$M_{btip} = 0.7F_{op} \times d_{tip} \tag{21}$$

The total bending moment occurring during an earthquake can be calculated using Equation (22).

$$M = M_{bs} + M_{bw} + M_{btip} \tag{22}$$

Next, the bending moment failure value of the critical section of the hollow composite insulator used with the bushing is calculated using Equations (23) and (24).

$$M_D = Z \times \sigma \tag{23}$$

$$Z = \pi(D^4 - d^4)/(32D) \tag{24}$$

In the equations, Z represents the flexural modulus of the porcelain outer sleeve, Mpa; D is the outer diameter at the root of the porcelain outer sleeve, m; d is the inner diameter at the root of the porcelain outer sleeve, m; σ denotes the flexural failure stress of the porcelain material, Mpa. If the calculated bending moment $M < M_D$, it is concluded that the bushing can withstand the seismic intensity corresponding to the given earthquake magnitude.

2.2.3. A Mathematical Model for Thermal Short-Circuit Current

According to GB/T 4109-2022 Clause 8.9.3, if the θ_f does not exceed 180 °C, the bushing is considered capable of withstanding the standard thermal short-circuit current I_{th} [19]. If θ_f is less than 180 °C, the thermal short-circuit current withstand test may be exempt. The capability of the bushing to withstand the I_{th} standard is calculated using Equation (25):

$$\theta_f = \theta_0 + \alpha \frac{I_{th}^2}{S_t \times S_e} \times t_{th} \tag{25}$$

where θ_f is the final temperature of the conductor, °C; θ_0 is the conductor temperature under continuous operation at the current I_r with an ambient temperature of 40 °C,°C; α is a coefficient, equal to 0.8(k/s)/(kA/cm^2)^2 for copper and 1.8(k/s)/(kA/cm^2)^2 for aluminum; t_{th} is the specified rated duration, s; I_{th} is the standard value specified above, kA; S_t is the total cross-sectional area corresponding to I_r, cm^2; S_e is the equivalent cross-sectional area considering the skin effect, cm^2, and is calculated using Equation (26):

$$S_e = \pi d(D - d) \tag{26}$$

where D is the diameter of the circular cross-section conductor, cm; d is the penetration depth of the current, cm, and is calculated using Equation (27):

$$d = \frac{1}{2\pi}\sqrt{\frac{\rho \times 10^3}{f}} \quad (27)$$

where ρ is the resistivity, μΩ·cm; f is the frequency, set to 50 Hz. If the calculated θ_f is less than 180 °C, the bushing is considered capable of withstanding the standard value, I_{th}.

2.2.4. A Mathematical Model for Thermal Field Simulation

The heat transfer in the bushings usually occurs in three forms: heat conduction, heat convection, and heat radiation. Since the heat transferred through radiation in the bushing is relatively small, radiation heat transfer is ignored in the simulation study. The solid-side heat conduction equation is as follows [20]:

$$\frac{k}{r}\frac{\partial}{\partial r}\left(r\frac{\partial T_c}{\partial r}\right) + k\frac{\partial^2 T_c}{\partial Z^2} + Q_c = 0 \quad (28)$$

where T_c represents the temperature of the core material, °C; k is the thermal conductivity; Q_c is the heat flux of the core material, J/(m^2·s). The heat transfer equation for convective heat transfer on the gas side is as follows:

$$Q_g = hA(T_c - T_g) \quad (29)$$

where Q_g represents the convective heat transfer flux between the bushing surface and the gas, W; h is the convective heat transfer coefficient; A is the effective area of the gas–solid interface, m^2; T_g is the gas temperature at the solid surface, °C. The governing equations considering both heat conduction and heat convection are as follows:

$$\rho c_p u \cdot \nabla T_g = \nabla \cdot (k \nabla T_g) + Q_g(\nabla \cdot (\rho u) = 0) \quad (30)$$

where ρ represents the density of the gas, kg/m^3; c_p is the specific heat capacity of the gas at constant pressure, J/(kg·K); u represents the velocity of the gas molecules, m/s.

2.3. Model Parameters and Boundary Conditions

The physical fields set in the valve side bushing include alternating current field, solid heat transfer field, and laminar flow field. The boundary conditions of the bushing simulation model are shown in Table 2, and the parameter settings are shown in Table 3.

Table 2. Simulation boundary conditions for bushings of different voltage levels.

Modules	Boundary Conditions	Unit
Conductive rod, 0 # screen	high voltage	V
Other screens	floating potential	V
Bottom screen, flange	grounding	V
Air	25	°C
Oil	85	°C
Heat transfer coefficient	5	W/(m^2·K)
Current of conductive rod	2000 (72.5 kV) 1250 (126 kV)	A

Table 3. Material parameters of bushings for the simulation calculation.

Materials	Density/ (kg/m^3)	Heat Conductivity/ [W/(m·K)]	Constant Pressure Heat Capacity/[J/(kg·K)]	Conductivity/ (S/m)	Relative Permittivity
Copper	8.96	400	385	5.998 × 107	1
Epoxy resin-impregnated capacitor core	2.21	0.10	730	7 × 10−15	4.2
Aluminum foil screen	3.90	237	900	1 × 108	1
Silicone rubber	2.20	0.20	703	1 × 10−14	3.75
Air domain	1.29	0.25	1003	1 × 10−18	1

2.4. Bushing Performance Test Methods

Based on the bushing structure described herein, bushings of different voltage levels were manufactured. Testing was conducted in accordance with IEC 60137-2017 [21] for power frequency withstand-voltage test, power-frequency wet withstand-voltage test, lightning impulse full-wave test, partial discharge test, and temperature rise test. A seismic test was performed in compliance with GB/T 50260 [22]. Testing encompassed destructive and type tests. The power frequency withstand test was destructive, with voltage applied continuously until breakdown occurred. The power frequency wet withstand test required 42.5/72.5/126 kV bushings to withstand 95/141/230 kV voltages for 60 s to qualify as compliant. The temperature rise test required 42.5/72.5/126 kV bushings to withstand 1250 A current for 5 h with a 1 h stabilization period. Oil temperature was measured post-testing completion. All tests were conducted at the Testing Center of Nanjing Electrical Science and Technology Group Co., Ltd. (Nanjing, China).

3. Results and Discussion

3.1. Simulation of Electric Field Strength in the Epoxy Core

Epoxy resin paper-impregnated bushings operate under high-voltage conditions, with concentrated electric fields at the high-voltage terminals, which can lead to insulation issues over long-term operation. Increasing the outer diameter of the capacitor core and the number of capacitor layers are common methods to optimize electric field distribution and enhance insulation performance. However, increasing the capacitor core size can lead to higher stress concentrations during curing, reduced heat dissipation performance, and increased manufacturing costs [23,24]. In this study, a design method with equal thickness and unequal steps is adopted to optimize the electric field distribution of the bushing. By adjusting the umbrella-step difference and tail-step difference to control the length of the steps, the radial and axial electric field strengths of the bushing are regulated, aiming to make the radial field strength as uniform as possible. Simultaneously, the axial field strength is controlled to be below the allowable axial field strength, ensuring sufficient insulation design margin. The threshold of radial and axial electric field strength is 10 kV/mm, which is determined by the breakdown strength of the epoxy resin.

In this study's design of the epoxy core, the outer diameter of the 40.5 kV bushing core is selected as 92 mm with 7 layers of capacitor screens; the outer diameter of the 72.5 kV bushing core is 116 mm with 8 layers of capacitor screens; and the outer diameter of the 126 kV bushing core is 130 mm with 10 layers of capacitor screens. The radial and axial electric field strengths of the epoxy core in the three different bushings were simulated and calculated, with the simulation results shown in Figure 4.

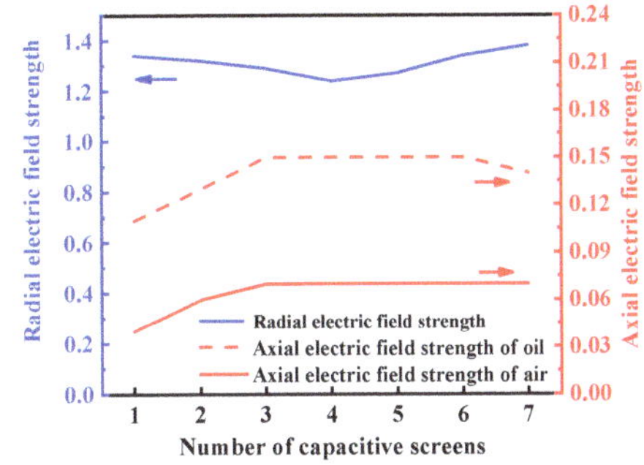

(a) Electric field strength distribution of a 40.5 kV bushing

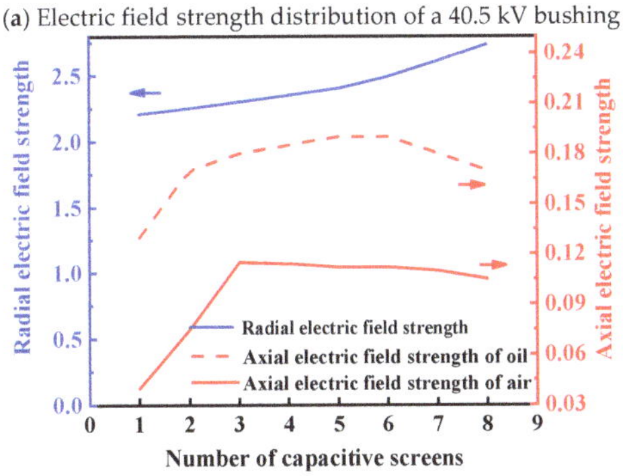

(b) Electric field strength distribution of a 72.5 kV bushing

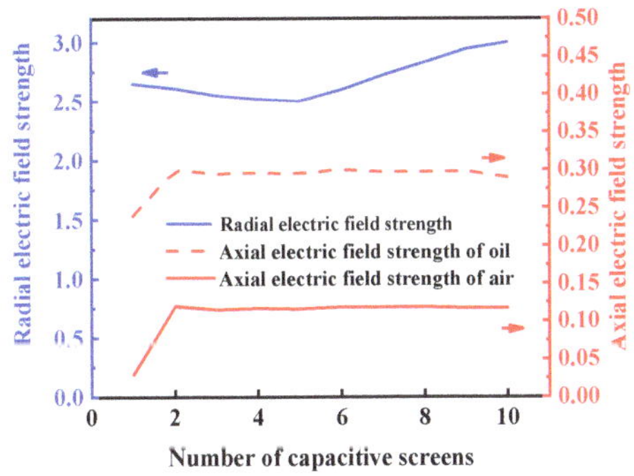

(c) Electric field strength distribution of a 126 kV bushing

Figure 4. Simulation results of electric field strength in the bushing core for different voltage levels. (The blue arrow represents the ordinate of the blue curve, and the red arrow represents the ordinate of the red curve.)

The epoxy cores of the 40.5 kV, 72.5 kV, and 126 kV bushings exhibit excellent distribution characteristics in terms of electric field strength and insulation performance, meeting design and operational requirements. Bushings of higher voltage levels have more layers of capacitor cores. The radial electric field strength of the epoxy cores in the 40.5 kV and 126 kV bushings first decreases and then increases, while that of the 72.5 kV bushing gradually rises. Specifically, the maximum radial electric field strength of the 40.5 kV bushing's epoxy core is 1.38 kV/mm, with a minimum of 1.24 kV/mm; for the 72.5 kV bushing, these values are 2.74 kV/mm and 2.21 kV/mm, respectively; and for the 126 kV bushing, they are 3.0 kV/mm and 2.5 kV/mm. In terms of axial electric field strength, the maximum values at the oil end and air end of the 40.5 kV bushing's epoxy core are 0.15 kV/mm and

0.07 kV/mm, respectively; for the 72.5 kV bushing, these are 0.19 kV/mm and 0.11 kV/mm; and for the 126 kV bushing, they are 0.3 kV/mm and 0.12 kV/mm. All these values are below the allowable thresholds, indicating uniform and stable electric field distribution. Overall, the epoxy cores of these bushings demonstrate outstanding performance in electric field strength control and insulation properties, effectively satisfying the operational demands of different voltage levels and ensuring the safety and stability of the equipment.

3.2. Bushing Electric Field Simulation

3.2.1. Simulation of the Bushing Electric Field Under Phase Voltage

In high-voltage bushings, due to the presence of capacitor screen structures between the core and the porcelain housing, the electric field distribution may not be uniform. Areas such as the edges of capacitor screens or the surface of the porcelain housing are prone to localized high field strengths. Before actual production, simulation calculations can be used to verify the rationality of the design, avoiding test failures or equipment damage caused by design flaws, thereby reducing development costs and time. The overall electric field distribution characteristics of 72.5 kV and 126 kV bushings under phase voltage were calculated.

Figure 5 shows the overall potential and radial electric field strength distribution of the 72.5 kV bushing under a phase voltage of 41.86 kV. Figure 6 shows the overall potential and radial electric field strength distribution of the 126 kV bushing under a phase voltage of 72.75 kV. It can be observed that, under a phase voltage of 41.86 kV, the highest electric field strength occurs at the first capacitor screen, reaching 2.76 kV/mm, while the lowest electric field strength occurs at the last screen, at 2.32 kV/mm. Under a phase voltage of 72.75 kV, the highest electric field strength at the first capacitor screen is 3.02 kV/mm, while the electric field strength at the last screen is 2.77 kV/mm, with the lowest electric field strength at the intermediate screens being 2.43 kV/mm. The trend of electric field strength distribution across the entire capacitor screen matches the calculated distribution curve.

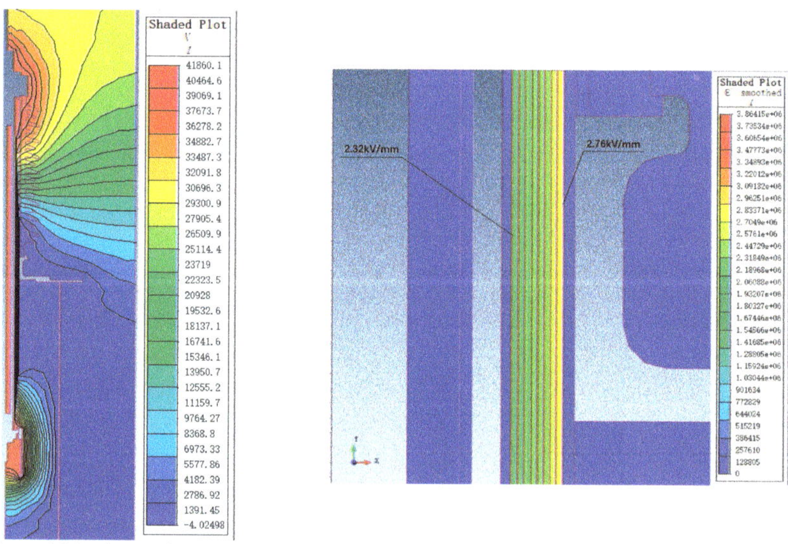

(**a**) Overall potential distribution of the 72.5 kV bushing(**b**) The radial electric field distribution diagram of the 72.5 kV bushing

Figure 5. The electric field distribution diagram of the 72.5 kV bushing under the 41.86 kV phase voltage.

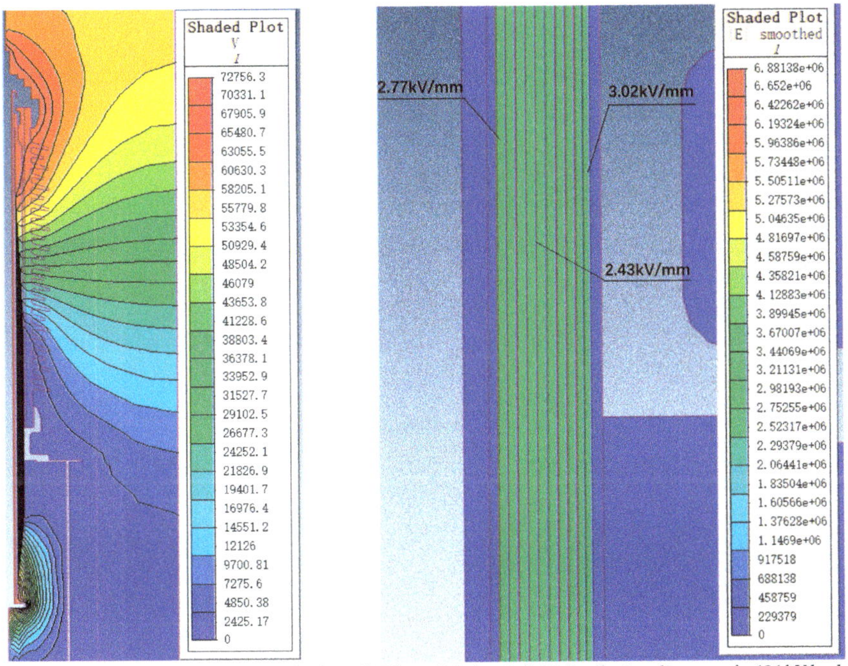

(**a**) Overall potential distribution of a 126 kV bushing(**b**) The radial electric field distribution diagram of a 126 kV bushing

Figure 6. The electric field distribution diagram of a 126 kV bushing under 72.75 kV phase voltage.

3.2.2. Simulation of the Bushing Terminal Electric Field Under Power Frequency Voltage

Using the finite element simulation method, the terminal electric field simulation of 72.5 kV and 126 kV bushings under power frequency voltage was conducted. For the 72.5 kV and 126 kV bushings, we applied 155 kV and 255 kV power frequency voltages, respectively, to calculate their electric field distributions. These voltage levels were selected in accordance with IEC 60137:2017.

Figures 7 and 8 show the electric field distribution diagrams of the air end and transformer end of the 72.5 kV bushing under a power frequency voltage of 155 kV. Figures 9 and 10, respectively, display the electric field distribution diagrams of the air end and transformer end of the 126 kV bushing under a power frequency voltage of 255 kV.

Under 155 kV power frequency voltage, the maximum electric field strength on the metal components at the air end of the bushing is 1.08 kV/mm, while the maximum electric field strength on the porcelain housing surface is 0.70 kV/mm. The electric field distribution at the air end is relatively uniform, with all field strength values within a reasonable range. At the transformer end of the bushing, the maximum electric field strength in oil is 5.14 kV/mm, located at the lower arc of the voltage balancing sphere. This area experiences a higher local field strength due to the radius of a curvature effect at the oil–sphere interface, yet it remains within the allowable limits. The maximum electric field strength on the core surface is 1.47 kV/mm, with the maximum tangential electric field strength on the core surface being 1.22 kV/mm, both below the allowable field strength thresholds. Simulation results indicate that the field distribution within the core is uniform, with all field strength values below the allowable standards. In conclusion, the electric field distribution of the 72.5 kV bushing designed in this study under 155 kV power frequency voltage is reasonable, with all local field strength values meeting relevant standard requirements.

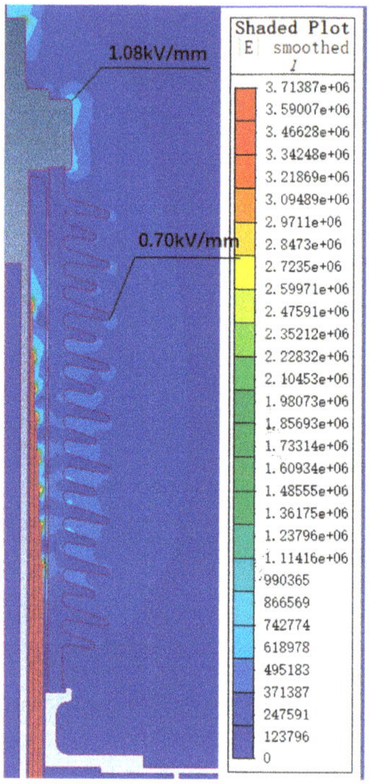

Figure 7. The electric field distribution diagram of a 72.5 kV bushing air end under 155 kV power frequency voltage.

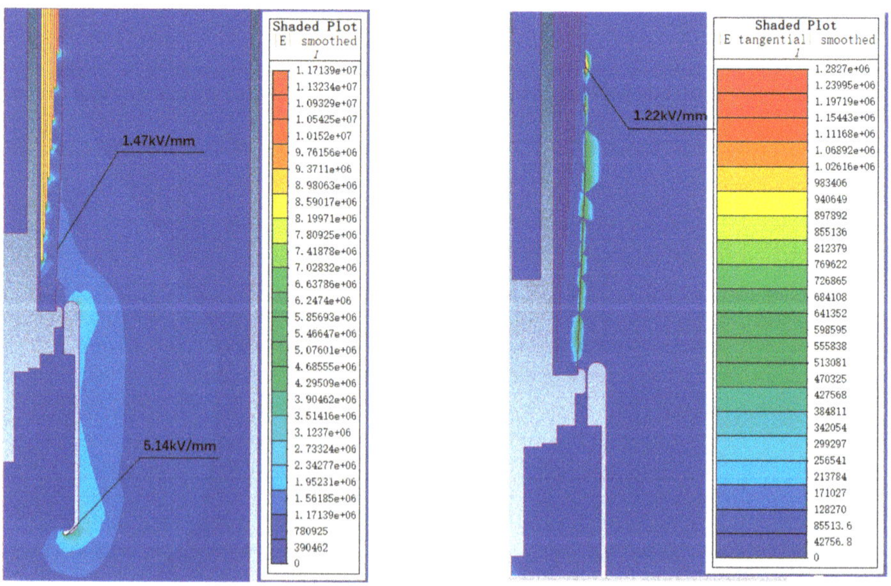

(**a**) Surface electric field strength (**b**) Tangential electric field strength

Figure 8. The electric field distribution diagram of a 72.5 kV bushing transformer end under 155 kV power frequency voltage.

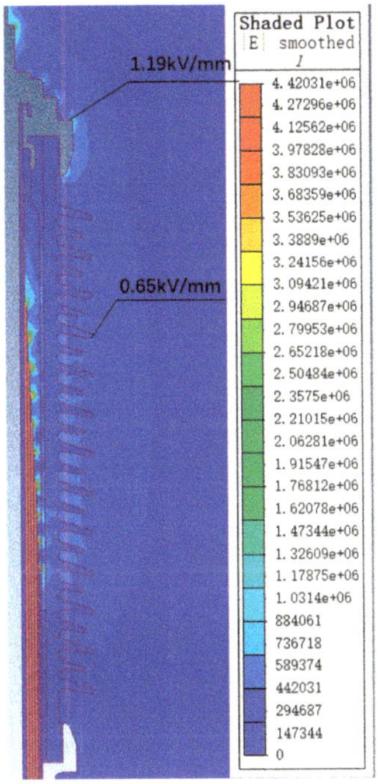

Figure 9. The electric field distribution diagram of a 126 kV bushing air end under 255 kV power frequency voltage.

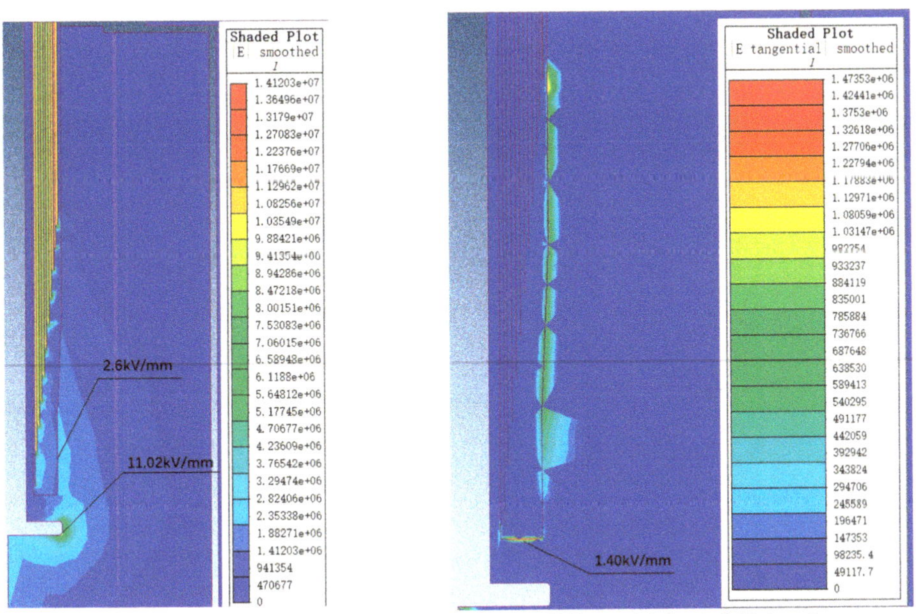

(a) Surface electric field strength (b) Tangential electric field strength

Figure 10. The electric field distribution diagram of a 126 kV bushing transformer end under 255 kV power frequency voltage.

Under 255 kV power frequency voltage, the maximum electric field strength on the metal components at the air end of the bushing is 1.19 kV/mm, while the maximum electric field strength on the porcelain housing surface is 0.65 kV/mm. At the transformer end of the bushing, the maximum electric field strength in oil is 11.02 kV/mm, located on the surface of the oil-immersed metal components. This area experiences a higher local field strength due to the field concentration effect at the metal surface, yet it remains within the allowable limits specified by relevant standards. The maximum electric field strength on the core surface is 2.6 kV/mm, with the maximum tangential electric field strength on the core surface being 1.40 kV/mm. Simulation results indicate that the electric field distribution within the core is uniform, with all field strength values below the allowable field strength standards.

Simulation results demonstrate that the field strengths at critical locations meet design requirements under power frequency voltage. This ensures the bushing's insulation performance and reliability during operation. Through finite element simulation analysis, the law of internal electric field distribution within the bushing has been revealed, providing a theoretical basis for optimizing the bushing design and enhancing its insulation performance.

3.3. Seismic Performance Calculation

In actual operation, bushings serve as critical support equipment in power systems and may be influenced by various external factors, including earthquakes. Earthquakes, as unpredictable natural disasters, are characterized by their suddenness and destructive power, potentially causing structural damage, insulation failure, and even severe safety incidents. Therefore, evaluating the seismic performance of critical equipment such as high-voltage bushings is a crucial aspect of ensuring the safe operation of power systems.

According to the product design specifications, bushings with voltage ratings of 40.5 kV, 72.5 kV, and 126 kV must withstand an 8-degree seismic intensity. The selection of calculation parameters for different products is presented in Table 4.

Table 4. Seismic calculation parameters for bushings of different voltage levels.

Voltage Level (kV)	40.5	72.5	126
Seismic intensity (degrees)	8	8	8
a_{gh}	0.3 g	0.3 g	0.3 g
a_{gv}	0.15 g	0.15 g	0.15 g
m_p (kg)	78	130	220
d_p (mm)	386	402	585
A (°)	0	0	0
P (Pa)	70	70	70
D_e (mm)	270	296	350
D_i (mm)	230	256	310
d_h (mm)	695	931	1530
d_b (mm)	498	506	666
F_{op} (N)	2000	2000	4000
d_{tip} (mm)	880	1156	1625
D (m)	0.152	0.168	0.2
D (m)	0.105	0.125	0.148
σ (MPa)	50	50	50

The results obtained based on the above parameters are shown in Table 5.

Table 5. Seismic calculation results for bushings of different voltage levels.

Voltage Level (kV)	40.5	72.5	126
M_{bs} (kN·m)	0.66	1.15	2.84
M_{bw} (kN·m)	0.06	0.09	0.24
M_{btip} (kN·m)	1.23	1.62	4.55
M (kN·m)	1.96	2.86	7.63
M_D (kN·m)	13.5	16	27.5

The 40.5 kV bushing, due to its relatively low voltage rating, has a smaller size and lighter mass, resulting in a simpler center of gravity position and force distribution. Under seismic loading, the total bending moment is 1.96 kN·m, which is significantly lower than the bending moment failure value (13.5 kN·m) at the critical section of the porcelain housing. This indicates that bushings with lower voltage ratings have a high safety margin in their design. As the voltage level of the bushing increases, its size and weight also increase, which makes the force distribution more complex. The total bending moments for the 72.5 kV and 126 kV bushings are 2.86 kN·m and 7.63 kN·m, respectively, which are still far below the bending moment failure values at the critical sections of their respective porcelain housings. Based on the calculation results, the total bending moments of the bushings under seismic loading are all significantly lower than the bending moment failure values at the critical sections of their porcelain housings, effectively satisfying the seismic intensity requirements of the design.

3.4. Thermal Short-Circuit Current Calculation of Bushings

In power systems, short-circuit faults are a common form of electrical failure, characterized by a sudden and significant increase in current, which may lead to equipment overheating or even damage. Therefore, the calculation of thermal short-circuit current (TSCC) is an essential task in the design and verification of power equipment, primarily used to evaluate the thermal withstand capability of the equipment under short-circuit fault conditions.

The selection of calculation parameters for different products is shown in Table 6. The results obtained based on the above parameters are shown in Table 7.

Table 6. Thermal short-circuit current calculation parameters for bushings of different voltage levels.

Voltage Level (kV)	40.5	72.5	126
θ_0 (°C)	120	120	120
I_{th} (kA)	31.25	50	31.25
t_{th} (s)	2	2	2
P (μΩ·cm)	7.1	1.85	2.83
D (cm)	4.5	6.5	6.5

Table 7. Thermal short-circuit current calculation results for bushings of different voltage levels.

Voltage Level (kV)	40.5	72.5	126
S_t (cm^2)	15.9	22.96	17.27
d (cm)	1.895	0.969	1.198
S_e (cm^2)	15.5	16.8	17.3
θ_f (°C)	126	130	131

For the 40.5 kV bushing, the cross-sectional area of the conductor corresponding to its designed rated current is 15.9 cm^2. Under the action of short-circuit current, due to the relatively small cross-sectional area, the current density carried per unit area is relatively high. This led to a rapid increase in the temperature of the conductor, eventually reaching 126 °C, but it was still lower than the 180 °C safety threshold stipulated by the national standard. With the voltage level increasing to 72.5 kV and 126 kV, in order to meet the demand for higher rated current transmission, the cross-sectional area of the conductor correspondingly increased to 22.96 cm^2 and 17.27 cm^2. The larger cross-sectional area effectively reduced the current density during short circuits, enabling the 72.5 kV and 126 kV bushings to significantly reduce the temperature rise of the conductors when subjected to greater short-circuit currents. Eventually, the temperatures were stabilized at 130 °C and 131 °C, respectively. Through systematic calculation and verification, the final conductor temperatures of the bushings of each voltage level under the thermal short-circuit condition are all lower than the limit values stipulated by the national standards. They can effectively withstand the standard short-time withstand current I_{th} and fully comply with the requirements of the safety operation specifications for power equipment.

3.5. Bushing Thermal Field Simulation Calculation

Through thermal field simulation, the current-carrying designs of bushings at different voltage levels were verified. Simulation parameters were set based on the actual operating conditions of bushings at different voltage levels. For the 72.5 kV bushing, the oil temperature was set to 85 °C, the air end to 25 °C, and the current was set to 2000 A for the simulation. For the 126 kV bushing, the oil temperature was set to 85 °C, the air end to 25 °C, and the current was set to 1250 A for the simulation. The maximum temperature of the conductor was 120 °C, which is stipulated by the standard GB/T 4109-2022 [19]. The overall temperature distributions for the 72.5 kV and 126 kV bushings are shown in Figures 11 and 12, respectively.

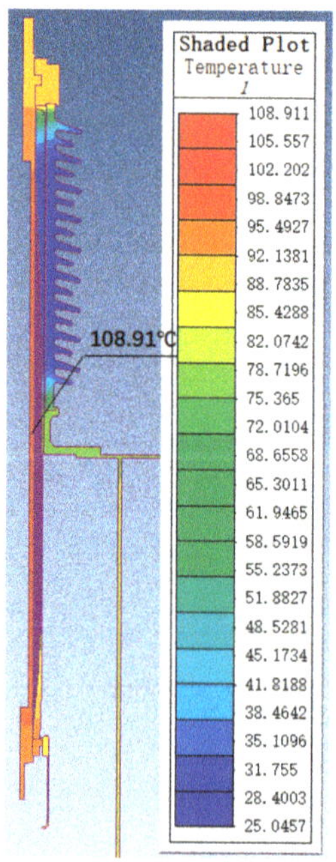

Figure 11. Temperature distribution of a 72.5 kV bushing under 2000 A current.

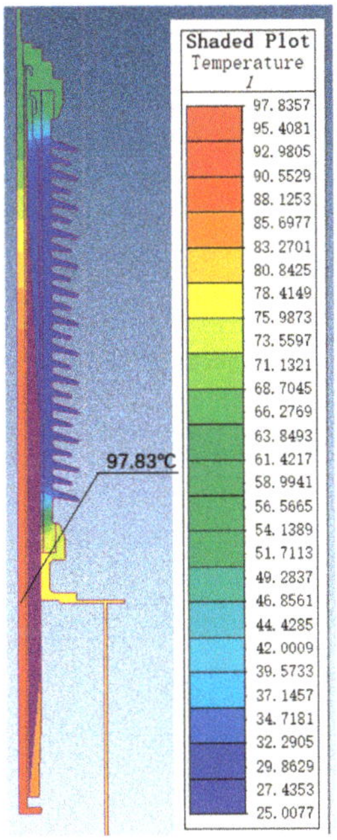

Figure 12. Temperature distribution of a 126 kV bushing under 1250 A current.

The uniformity of the electric field distribution directly affects the electrothermal conversion efficiency within the bushing. In regions with higher electric field strength, the electrothermal conversion efficiency is higher, potentially leading to localized temperature increases. The maximum radial electric field strengths for the 72.5 kV and 126 kV bushings are 2.74 kV/mm and 3.0 kV/mm, respectively, both within permissible limits. The corresponding thermal field simulation results indicate that the maximum temperature for the 72.5 kV bushing is 109 °C, located in the middle of the conductor, while the maximum temperature for the 126 kV bushing is 98 °C, located in the middle-lower part of the conductor. The calculation results demonstrate that bushings at higher voltage levels exhibit excellent heat dissipation performance in their designs, with uniform temperature distributions, ensuring thermal stability under high voltage and high current conditions. The temperature distribution of the bushing under actual operating conditions also supports the results of thermal short-circuit current calculations.

The uniformity of the electric field distribution, structural integrity, and mechanical stability collectively ensure the thermal stability of the bushing. Uniform electric field distribution reduces the risk of localized overheating, structural integrity ensures the uniformity of the thermal field distribution, and mechanical stability guarantees the reliability of the bushing under extreme operating conditions.

3.6. Bushing Performance Testing

We conducted the necessary tests for the operation of bushings of different voltage levels. Testing results are presented in Table 8. Test results indicate that bushings of different voltage classes, manufactured based on the structure described herein, meet standard requirements for insulation, heat resistance, and seismic performance. The 42.5/72.5/125 kV bushings can withstand 105/165/305 kV power frequency voltages, significantly exceeding their maximum operating voltages. Moreover, their power frequency wet withstand

and lightning impulse full-wave tests were fully compliant, indicating excellent outdoor performance, effectively withstanding inevitable rain and lightning impacts. Results of temperature rise and seismic tests show that bushings possess superior heat resistance and seismic performance. These findings provide a reliable theoretical basis for the development and application of epoxy resin paper-insulated bushings, ensuring reliability under high voltage classes and complex conditions. This research holds a significant engineering value for enhancing the safety and stability of transformers and power systems.

Table 8. Performance testing results of bushings at different voltage levels.

Voltage Level (kV)	Power Frequency Withstand-Voltage Test (kV)	Power Frequency Wet Withstand-Voltage Test (s)	Lightning Impulse Full-Wave Test (kV)	Partial Discharge Test (pC)	Temperature Rise Test (°C)	Seismic Test
40.5	105	60	250	1	75.8	0.3 g
72.5	165	60	380	1	74.5	0.3 g
126	305	60	650	1	72.4	0.3 g

4. Conclusions

As a critical component of transformer systems, epoxy resin paper-impregnated bushings play a vital role in ensuring the system's reliable operation. Their electric field distribution, seismic resistance, and thermal stability are key factors determining the reliable operation of the transformer system. This study employs finite element simulation to analyze the electric field distribution characteristics of 40.5 kV, 72.5 kV, and 126 kV bushings. The mathematical model of the bushing is constructed to calculate and predict the seismic resistance and thermal stability of the bushing under working conditions. The research provides a reliable method for the development and design of high-reliability high-voltage direct current bushings. The relevant results are as follows:

1. The electric field strength distribution varies among bushings with different structures. The maximum radial electric field strengths for the 40.5 kV, 72.5 kV, and 126 kV bushings are 1.39 kV/mm, 2.86 kV/mm, and 3.0 kV/mm, respectively, all meeting the usage requirements. The maximum axial electric field strengths at both the oil end and air end of the bushing are below the allowable axial electric field strength, satisfying the usage requirements.
2. The bushings with rated voltages of 40.5 kV, 72.5 kV, and 126 kV meet the 8-degree seismic resistance requirements and can effectively withstand the standard seismic intensity value I_{th}. They are also capable of withstanding the electromagnetic forces caused by dynamic stability currents, in compliance with standard specifications. The bushings demonstrate reliable operational capabilities under extreme scenarios.
3. The thermal stability of the bushings under short-circuit conditions complies with national standards. The final conductor temperature in all cases is below 180 °C, providing sufficient safety margins. The thermal field simulation results indicate that the maximum temperature of the 72.5 kV bushing under a current of 2000 A is 109 °C, and the maximum temperature of the 126 kV bushing under a current of 1250 A is 98 °C, both satisfying engineering usage requirements.

Author Contributions: Conceptualization, D.L. and L.W.; data curation, D.L., X.D., T.Y. and Z.L.; formal analysis, X.T. and T.Y.; funding acquisition, L.W.; investigation, L.L. and W.T.; methodology, B.C. and Z.L.; project administration, W.T. and L.W.; resources, D.L. and X.T.; supervision, L.L. and B.C.; validation, X.D.; writing—original draft, Z.L.; writing—review and editing, B.C. All authors have read and agreed to the published version of the manuscript.

Funding: This work was funded by the China Yangtze Power Co., Ltd. Scientific Research Project Grant No. 2322020029, entitled "Research on independent and controllable technology of large-scale main transformer bushing".

Institutional Review Board Statement: Not applicable.

Data Availability Statement: The original contributions presented in this study are included in the article. Further inquiries can be directed to the corresponding authors.

Conflicts of Interest: Authors Daijun Liu, Xiaoying Dong, and Tianming Yan were employed by China Yangtze Power Co., Ltd. Authors Xiaobang Tong, Libao Liu, and Wenkai Tang were employed by Nanjing Electric High Voltage Bushing Co., Ltd. The remaining authors declare that the research was conducted in the absence of any commercial or financial relationships that could be construed as a potential conflict of interest.

References

1. Zheng, H.; Li, Y.; Luo, X.; Zhang, E.; Jing, J. Investigation on Hot Spot Temperature of Resin-Impregnated Paper High-Voltage Bushing Based Upon Dielectric Loss. *IEEE Trans. Dielectr. Electr. Insul.* **2023**, *30*, 1884–1892. [CrossRef]
2. Shi, S.; Tian, H.; Xie, G.; Wang, Q.; Xie, T.; Pang, X.; Yang, G.; Liu, P.; Peng, Z. Insulation Degradation of Epoxy Resin Impregnated Paper Bushing Under Discharge Arc Ablation. *IEEE Trans. Dielectr. Electr. Insul.* **2024**, *31*, 2776–2784. [CrossRef]
3. Shang, G.; Liu, X.; Shao, Y.; Chen, M.; Sun, Y.; Zhang, Q.; Tang, H.; Li, X. Insulation Degradation Characteristics and Phase-Field Simulation of Epoxy Resin for Valve-Side Bushing Under Strongly Coupled Electrothermal Field. *IEEE Trans. Dielectr. Electr. Insul.* **2024**, *31*, 615–624. [CrossRef]
4. Li, L.; Li, Q.; Xu, S.; Liu, R.; Dong, M.; Ying, S.; Tian, J.; Xin, W.; Haddad, M.; Jiang, X. Electric Field Improvement for High-Voltage Bushings. *Polymers* **2023**, *15*, 40. [CrossRef]
5. Zhang, H.; Jin, S.; Jin, H.; Zhang, Z.; Wang, C.; Liu, P. A novel simulation of space charge decay dynamics after removing the voltage in epoxy resin composites. *J. Phys. D Appl. Phys.* **2025**, *58*, 035303. [CrossRef]
6. Shang, G.; Sun, Y.; Tang, H.; Li, X.; Zhang, S.; Li, J.; Liu, X. Degradation mechanism for epoxy resins under combined electric, thermal and compressive stresses. *Polym. Degrad. Stab.* **2024**, *230*, 111038. [CrossRef]
7. Lin, M.; Wu, G.; Liu, K.; Yan, X.; Tang, H. Review of Detection Methods for Typical Faults in Transformer Bushings. *IEEE Electr. Insul. Mag.* **2024**, *40*, 33–44. [CrossRef]
8. Chen, M.; Liu, X.; Shao, Y.; Shang, G.; Zhang, Q.; Tang, H.; Liu, L. Degradation characteristics of insulation near aluminium foil edges inside dry-type bushing cores under electrothermal compound stress. *High Volt.* **2022**, *7*, 1153–1164. [CrossRef]
9. Sun, H.; Yang, D.; Shi, W.; Sun, Y.; Liu, X. Influence of Multi-Order Harmonic Voltage on Degradation in Epoxy Resin Layer of High Voltage Valve Side Bushing. *IEEE Trans. Dielectr. Electr. Insul.* **2025**, 3542002. [CrossRef]
10. Elsadd, M.A.; Elsad, R.A.; Huwayz, M.A.; Mansour, S.A.; Zaky, M.S.; Elkalashy, N.I.; Izzularab, M.A. Improving the distribution system capability by incorporating ZnO nanoparticles into high-density polyethylene cable materials. *Sci. Rep.* **2024**, *14*, 25834. [CrossRef] [PubMed]
11. Wang, Q.; Liao, J.; Tian, H.; Liu, P.; Peng, Z. Regularity analysis of the temperature distribution of epoxy impregnated paper converter transformer bushings. *IEEE Trans. Dielectr. Electr. Insul.* **2017**, *24*, 3254–3264. [CrossRef]
12. Wang, Q.; Wang, H.; Peng, Z.; Liu, P.; Zhang, T.; Hu, W. 3-D coupled electromagnetic-fluid-thermal analysis of epoxy impregnated paper converter transformer bushings. *IEEE Trans. Dielectr. Electr. Insul.* **2017**, *24*, 630–638. [CrossRef]
13. Yadav, S.; Chamorro, H.R.; Flores, W.C.; Mehta, R.K. Investigation of Improved Thermal Dissipation of ±800 kV Converter Transformer Bushing Employing Nano-Hexagonal Boron Nitride Paper Using FEM. *IEEE Access* **2021**, *9*, 149196–149217. [CrossRef]
14. Xu, S.; Liu, J.; Li, K.; Ma, H.; Chen, P.; Huang, K. Analysis of Transformer GSB Bushing Fault Causes Based on Finite Element Simulation. *J. Phys. Conf. Ser.* **2024**, *2800*, 012002. [CrossRef]
15. Zhu, W.; Xie, Q.; Lu, J. Seismic testing and fragility analysis considering material strength uncertainty of 1100 kV GFRP composite power transformer bushing. *Case Stud. Constr. Mater.* **2024**, *21*, e03564. [CrossRef]
16. He, C.; Xie, Q.; Zhou, Y. Influence of flange on seismic performance of 1100-kV ultra-high voltage transformer bushing. *Earthq. Spectra* **2019**, *35*, 447–469. [CrossRef]
17. Shi, Y.; Xie, G.; Wang, Q.; Li, X.; Yang, X.; Liu, P.; Peng, Z. Simulation analysis and calculation of electric field distribution characteristics of UHV wall bushing. *Energy Rep.* **2021**, *7*, 110–117. [CrossRef]
18. *IEC TS 61463*; Bushings-Seismic Qualification. International Electrotechnical Commission: Geneva, Switzerland, 2016.
19. *GB/T 4109-2022*; Insulated Bushings for Alternating Voltages Above 1000 V. Standards Press of China: Beijing, China, 2022.
20. Lan, B.; Guo, N.; Li, J.; Tang, H.; Wang, X.; Ding, W. Multi-physical Field Coupling Simulation of DC Bushing Core Considering the Anisotropic Volume Conductivity of Resin Impregnated Paper. *High Volt. Eng.* **2025**, *40*, 900–912.

21. *IEC 60137:2017*; Insulated Bushings for Alternating Voltages Above 1000 V. International Electrotechnical Commission: Geneva Switzerland, 2017.
22. *GB/T 50260-2013*; Code for Seismic Design of Electrical Installations. Standards Press of China: Beijing, China, 2013.
23. Allahbakhshi, M.; Akbari, M. Heat analysis of the power transformer bushings using the finite element method. *Appl. Therm. Eng.* **2016**, *100*, 714–720. [CrossRef]
24. Sitar, R.; Janic, Z.; Stih, Z. Improvement of thermal performance of generator step-up transformers. *Appl. Therm. Eng.* **2015**, *78*, 516–524. [CrossRef]

energies

MDPI

Article

Influence of Ion Flow Field on the Design of Hybrid HVAC and HVDC Transmission Lines with Different Configurations

Jinyuan Xing [1], Chenze Han [2,*], Jun Tian [1], Hao Wu [1] and Tiebing Lu [2]

1 PowerChina Sichuan Electric Power Engineering Co., Ltd., Chengdu 610041, China; xingjinyuan90@163.com (J.X.); 13808078819@163.com (J.T.); hao_hao2006@163.com (H.W.)
2 School of Electrical and Electronic Engineering, North China Electric Power University, Beijing 102206, China; tiebinglu@ncepu.edu.cn
* Correspondence: hcz131028@163.com

Abstract: Due to the coupling of DC and AC components, the ion flow field of HVDC and HVAC transmission lines in the same corridor or even the same tower is complex and time-dependent. In order to effectively analyze the ground-level electric field of hybrid transmission lines, the Krylov subspace methods with pre-conditioning treatment are used to solve the discretization equations. By optimizing the coefficient matrix, the calculation efficiency of the iterative process of the electric field in the time domain is greatly increased. Based on the limit of electric field, radio interference and audible noise applied in China, the main factor influencing the design of hybrid transmission lines is determined in terms of electromagnetic environment. After the ground-level electric field of transmission lines with different configurations is analyzed, the minimum height and corridor width of double-circuit 500 kV HVAC lines and one-circuit ±800 kV HVDC lines in the same corridor are obtained. The research provides valuable practical recommendations for optimal tower configurations, minimum heights, and corridor widths under various electromagnetic constraints.

Keywords: electric field; HVDC transmission lines; hybrid transmission lines; ion flow field; numerical method

1. Introduction

For power transfer over long distances, HVDC transmission lines are more effective than HVAC transmission lines [1]. Therefore, HVDC overhead lines have been designed and installed in China, Brazil, India, Pakistan, and other countries in recent years. Since 2009, 20 ultra-HVDC overhead lines (voltage including ±800 kV and ±1100 kV) with a total length longer than 35,000 km have been in operation in China. In order to increase the power transmission capacity, many countries are also considering the conversion of AC lines into DC lines [2]. Consequently, the parallel HVAC and HVDC power transmission lines (also called hybrid transmission lines) will be common in the future.

There are two kinds of hybrid transmission lines: HVAC and HVDC lines on one tower, and parallel HVAC and HVDC lines on different towers. Because the electric field at HVDC conductor surface is changed with the power frequency cycle under the influence of HVAC lines, the corona discharge of parallel HVDC conductors is changed, and then space charges distribution is different from HVDC lines without parallel HVAC lines, which is illustrated in Figure 1. A, B, and C represent the three-phase HVAC conductors; "+" and "−" represent the positive and negative HVDC poles. A 2D plane perpendicular to the line shows ion flow trajectories and the ground electric field along segment ab from the DC lines is shown, which clearly manifests the influence of the AC lines. Due to the mutual influence of parallel AC and DC lines [3,4], with challenges posed for fault analysis and corona analysis of the lines additionally [5–7], the ion flow field becomes more complicated, which is the superposition of electric field generated by conductor surface charges and space charges due to the corona discharge [8].

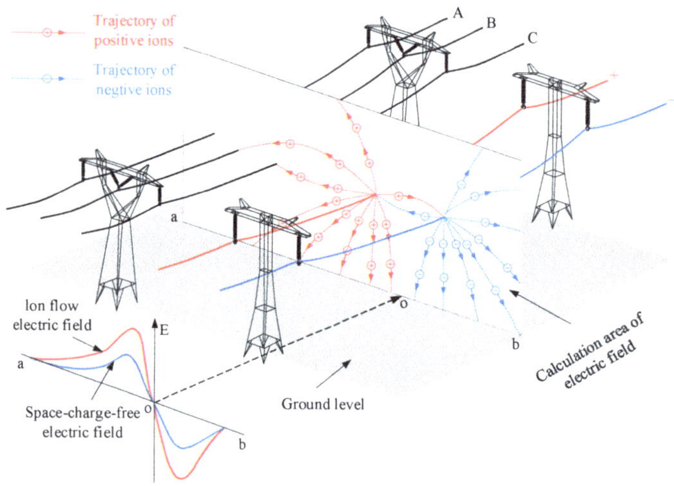

Figure 1. Electric field and space charges of hybrid AC and DC transmission lines on different towers.

In 2022, the transmission and distribution committee of IEEE Power and Energy Society published IEEE standard 2819 on measuring method of electromagnetic environment for the corridor of hybrid transmission lines [9], which can provide guidance for the monitoring of the ground-level electric field in the same corridor. At the same time, the limit value of the ground-level electric field to control the height and corridor width is proposed through human perception experiments [10]. But for the design of hybrid transmission lines, the mutual influence between parallel AC and DC lines on the ground-level electric field must be analyzed on the basis of the numerical calculation in order to obtain a suitable height and corridor width.

Chartier reported the test results of Bonneville power administration and analyzed characteristics of conductor surface gradient in 1981, but there was no calculation method to predict the ground-level electric field [8]. Since then, the effective calculation method has been an important research hotspot. In 1988, Maruvada first calculated the electric field of hybrid transmission lines by using the flux tracing method, which is based on the Deutsch assumption [11]. In 1989, Clairmont adopted the concept of corona saturation degree to analyze the electric field variation of hybrid lines in different configurations [12]. In 1990, Abdel-Salam calculated the electric field without consideration of the space charges' effect by using the charge simulation method [13]. In 1996, Zhao improved the flux tracing method by treating AC lines as several separate DC voltages during one power-frequency cycle [14], and in 2010, Yang further improved the same method in the time domain along the flux line based on the Deutsch assumption [15]. In 2009, Li proposed the time-domain upwind finite element method without the Deutsch assumption in order to analyze the electric field of hybrid 1000 kV AC and ±800 kV DC lines, and a variable time-step is applied to accelerate the computation process [16]. In 2011, Yin solved the electric field by the charge simulation method and the finite-element method (FEM), and solved the space charges by the time-dependent finite volume method (FVM) [17]. In 2012, Zhou proposed a time-efficient method based on FEM and FVM, where the significant effect of AC corona was analyzed [18]. In 2013, Straumann adopted the discontinuous Galerkin method to the solution of the electric field and discussed the simplification of more complex problems in order to obtain efficient computation [19]. In 2014, Guillod proposed the iterative method of characteristics similar to the flux tracing method, but the Deutsch assumption was abandoned [20]. In 2016, Zhang proposed the time-domain method of characteristics, and the electric field and ion density in the space and time domain were updated [21]. In 2017, Qiao proposed an upwind finite element method combining the domain decomposition technique with high-order elements, and introduced a conductor surface charge density updating strategy based on the corona discharge U–I characteristic curve [22]. Li applied this method to analyze the influence of four different AC conductor arrangements and phase sequence configurations in hybrid transmission lines on the distribution of ground-level electric field [23]. In 2018, Qiao proposed an iterative flux tracing method without the Deutsch assumption, which was more reasonable [24]. In 2018, Ma analyzed the 3D electric field of ±800 kV DC and 500 kV AC parallel lines based on the flux tracing method [25]. In 2019, Tian applied the time-domain mixed-hybrid finite element method to analyze the ion flow field of two typical hybrid transmission lines, improving accuracy and reducing computational burden [26]. In 2020, Xu successfully improved the solution efficiency based on a parallel processing algorithm with fine-grained nodal domain decomposition and an upwind nodal charge conservation method, and the electric field of hybrid lines on one tower was analyzed [27]. With the development of the above method, many characteristics have been obtained in order to understand the ground-level electric field under some special hybrid lines. But the design of hybrid lines, including the height, corridor width and the approach distance, is not emphasized in detail.

In this paper, FEM-FVM is used to solve the ion flow field of hybrid lines in different configurations. Section 2 details the FEM-FVM calculation methods and acceleration techniques. In order to increase the simulation efficiency, the conjugate gradient method with pre-condition treatment (PCG) is applied to treat the coefficient matrix in FEM discretization equation, and generalized minimal residual method (GMRES) with reverse Cuthill-McKee (RCM) and bipartite matching algorithm (BMA) is used to treat the FVM discretization solution. After the efficiency is compared, the ground-level electric field for transmission lines with different configurations is analyzed. Section 3 presents the simulation models and design constraints, focusing on conductor arrangements and electromagnetic limits in China. Finally, Sections 4 and 5 analyze the design of hybrid lines on one tower and different towers, respectively. The height and the corridor width of double-circuit HVAC lines and one-circuit HVDC lines in the same corridor are obtained, especially the effect of different configurations of AC lines on the design is discussed for hybrid lines on one tower and different towers. The study provides valuable practical recommendations for optimal tower configurations, minimum heights, and corridor widths under various electromagnetic constraints.

2. Calculation Method and Limits

In order to simulate the ion flow field of hybrid lines, the calculation method is introduced first.

2.1. Control Equations and FEM-FVM

The corona of hybrid lines generates a time-dependent bipolar ion flow field problem, which can be described by the following equations:

$$\nabla^2\varphi(t) = -\frac{\rho^+(t) - \rho^-(t)}{\varepsilon_0} \tag{1}$$

$$\frac{\partial\rho^{+}(t)}{\partial t}+\nabla\cdot\left[\rho^{+}(t)k^{+}\nabla\varphi(t)\right]=-\frac{R\rho^{+}(t)\rho^{-}(t)}{e} \tag{2}$$

$$\frac{\partial\rho^{-}(t)}{\partial t}-\nabla\cdot\left[\rho^{-}(t)k^{-}\nabla\varphi(t)\right]=-\frac{R\rho^{+}(t)\rho^{-}(t)}{e} \tag{3}$$

where φ is the electric potential in time domain, ρ is the charge density in time domain, the superscripts + and − are respectively for positive and negative charges, ε_0 is the dielectric constant of air, k is the ion mobility, $k^+ = 1.2 \times 10^{-4}$ m^2/V/s and $k^- = 1.5 \times 10^{-4}$ m^2/V/s, R is the ion recombination rate, and e is the electron charge. The influence of air electrical conductivity and air humidity is complicated for charging model establishment [28,29]. Therefore, both air electrical conductivity and humidity are neglected to simplify the model.

The potentials of the DC and AC conductors are set to the applied voltages U_{DC} and U_{AC}, while the ground potential is set to zero. The artificial boundary is defined as a semicircle with a radius six times the height of the transmission line, and the low charge density at this boundary has a negligible impact on the potential. The artificial boundary potential U_{Γ_3} is obtained using the charge simulation method. The boundary conditions of Poisson's equation are defined as follows:

$$\begin{cases} \varphi|_{\Gamma_1} = U_{DC} \\ \varphi|_{\Gamma_2} = U_{AC}(t) \\ \varphi|_{\Gamma_3} = U_{\Gamma_3}(t) \\ \varphi|_{\Gamma_4} = 0 \end{cases} \tag{4}$$

where Γ_1 is the boundary of the DC conductors, Γ_2 is the boundary of the AC conductors, Γ_3 is the artificial semicircular boundary, and Γ_4 is the ground boundary.

The boundary conditions of the current continuity equations are as follows. As the ionization region thickness around the conductor is neglected, the electric field strength on the conductor surface follows Kaptzov's assumption, whereby the surface electric field strength equals the corona onset field strength once the applied voltage exceeds the threshold. The conductor surface charge density is initially unknown and is iteratively adjusted at each time step using the secant method [18] to satisfy Kaptzov's assumption. The consideration of ion drift and the neglect of ion diffusion eliminates the need for boundary conditions on the current continuity equations at the ground and artificial boundaries. If charges flow into the control volume from the outside computational domain, the charge amount is set to zero. If charges flow out, no special treatment is applied.

Because of the coupling of potential and space charges, the iterative process in the time domain is necessary for the solution of the potential in (1) and the space charges in (2) and (3). FEM is applied to discretize Poisson's Equation (1), and FVM is used to discretize current continuity Equations (2) and (3).

After boundary conditions are applied, (1) can be changed to the following format in the whole calculation area by the treatment of FEM based on triangular meshes:

$$\mathbf{K}\boldsymbol{\varphi} = \mathbf{F} \tag{5}$$

where **K** is M*M sparse matrix which is determined by the FEM node relation, $\boldsymbol{\varphi}$ is the column vector of node potentials, **F** is the column vector including the effect of imposed boundary condition and node space charges, and M is the node number in the whole calculation area. The conjugate gradient with pre-conditioning treatment (PCG) is used to solve (5).

Based on the implicit time scheme, (2) can be changed as

$$\rho^+_{i,n+1} = \rho^+_{i,n} - \Delta t \cdot R\rho^+_{i,n+1}\rho^-_{i,n} - \frac{\Delta t}{A_i}\sum_{m=1}^{N_i} \rho^+_{i,n+1,m}k^+E_{i,n,m}l_{i,m} \tag{6}$$

where A_i, m and $\mathrm{N_i}$ are respectively the area, the edge number and the total number of edges in the i-th control volume; the subscript n and n + 1 are respectively for the n-th and n + 1-th time step number; Δt is the time step; $l_{i,m}$ is the length of the m-th edge in the i-th control volume; $\rho^+_{i,n+1,m}$ and $E_{i,n,m}$ are respectively the space charge and electric field in the center of the related edge. $E_{i,n,m}$ can be obtained from the upwind control volume using second-order interpolation:

$$\rho^+_{i,n+1,m} = \rho^+_{i,n+1,m,up} + \begin{bmatrix} x_{i,m} - x_{i,m,up,c} & y_{i,m} - y_{i,m,up,c} \end{bmatrix} \begin{bmatrix} \left(\frac{d\rho^+}{dx}\right)_{i,n+1,m,up} \\ \left(\frac{d\rho^+}{dy}\right)_{i,n+1,m,up} \end{bmatrix} \tag{7}$$

where $\rho^+_{i,n+1,\mathrm{m,up}}$ is the average charge density of the m-th edge of the i-th upwind control volume, $x_{i,m,up,c}$ and $y_{i,m,up,c}$ are the coordinates of the center of gravity in the upwind control volume, and the charge density gradient $(\nabla\rho^+)_{i,n+1,\mathrm{m,up}}$ can be obtained by

$$\begin{bmatrix} \Delta x_{\mathrm{i},m,up,1} & \Delta y_{\mathrm{i},m,up,1} \\ \Delta x_{\mathrm{i},m,up,2} & \Delta y_{\mathrm{i},m,up,2} \\ \vdots & \vdots \\ \Delta x_{\mathrm{i},m,up,N} & \Delta y_{\mathrm{i},m,up,N} \end{bmatrix} \begin{bmatrix} \left(\frac{d\rho^+}{dx}\right)_{\mathrm{i},n+1,m,up} \\ \left(\frac{d\rho^+}{dy}\right)_{\mathrm{i},n+1,m,up} \end{bmatrix} = \begin{bmatrix} \rho^+_{\mathrm{i},n+1,m,up,1} - \rho^+_{\mathrm{i},n+1,m,up} \\ \rho^+_{\mathrm{i},n+1,m,up,2} - \rho^+_{\mathrm{i},n+1,m,up} \\ \vdots \\ \rho^+_{\mathrm{i},n+1,m,up,N} - \rho^+_{\mathrm{i},n+1,m,up} \end{bmatrix} \tag{8}$$

where N is the number of control volumes adjacent to the upwind control volume. Equation (8) can be solved by using the least squares method. Based on (7) and (8), $\rho^+_{i,n+1,m}$ can be expressed as a polynomial in terms of $\rho^+_{i,n+1,m,up}, \rho^+_{i,n+1,m,up,1}, \rho^+_{i,n+1,m,up,2} \cdots \rho^+_{i,n+1,m,up,N}$. In the mesh with node N_0, Equation (6) can be expanded as:

$$\begin{cases} A_{11}\rho^+_{1,n+1} + A_{12}\rho^+_{2,n+1} + \ldots + A_{1m}\rho^+_{m,n+1} + \ldots + A_{1\mathrm{N_0}}\rho^+_{\mathrm{N_0},n+1} = \rho^+_{1,n} \\ A_{21}\rho^+_{1,n+1} + A_{22}\rho^+_{2,n+1} + \ldots + A_{2m}\rho^+_{m,n+1} + \ldots + A_{2\mathrm{N_0}}\rho^+_{\mathrm{N_0},n+1} = \rho^+_{2,n} \\ \ldots\ldots \\ A_{\mathrm{N_0}1}\rho^+_{1,n+1} + A_{\mathrm{N_0}2}\rho^+_{2,n+1} + \ldots + A_{\mathrm{N_0}m}\rho^+_{m,n+1} + \ldots + A_{\mathrm{N_0N_0}}\rho^+_{\mathrm{N_0},n+1} = \rho^+_{\mathrm{N_0},n} \end{cases} \tag{9}$$

Equation (3) can be changed to the similar format as (9) just instead of positive space charges with negative space charges. Then, (2) and (3) can be expressed in the following equations after the treatment of FVM:

$$\mathbf{A\rho} = \mathbf{B} \tag{10}$$

which is similar to (5), although each matrix has a different meaning. The coefficient matrix **A** is not symmetric due to the varying adjacency of each control volume and the asymmetric mutual influence. The GMRES method is used to solve the above equation.

Methods for calculating the hybrid electric field include the finite difference method (FDM) and the flux tracing method, the FEM–FVM and the time-domain upwind finite element method. FDM offers higher computational efficiency but has difficulties in modeling complex models on structured meshes, and local conservation of physical quantities cannot always be guaranteed during the discretization process, particularly in scenarios involving strong convection or sharp gradients. Additionally, the flux tracing method is an analytical solution under the Deutsch assumption, which provides an efficient framework by reducing 2D ion-flow field problems to 1D approximations. While it is computationally efficient and free from numerical oscillations, it fundamentally neglects the distortion of the electric field direction caused by space charge effects. Qiao [24] proposed an iterative flux tracing algorithm, which can progressively reduce Deutsch assumption errors by recursively recalculating space charge-modified electric fields and updating flux line trajectories. The FEM-FVM has two key features: FVM ensures flux conservation, and this method employs iterative computations at each time step to determine charge emission values satisfying the Kaptzov assumption. Both the time-domain upwind finite element method and the FEM-FVM are based on the FEM for solving Poisson's equation similarly. The time-domain upwind finite element method uses a time-dependent upwind first-order difference method on finite element meshes for the current continuity equations, while FEM-FVM uses a second-order finite volume method based on control volume meshes. In terms of temporal discretization, the time-domain upwind finite element method uses an explicit method, whereas FEM-FVM uses an implicit method. This leads to higher accuracy for FEM-FVM but results in a significant increase in computational load. Therefore, accelerating the FEM-FVM becomes important [30].

2.2. Acceleration Method

According to the previous discussion, Poisson's equation and current continuity equations can essentially be changed to matrix Equations (5) and (10) after the treatment of FEM-FVM discretization. Because the iteration solution of (5) and (10) in time domain requires significant memory and computational effort, the solution including PCG for FEM discretization and GMRES for FVM discretization is necessary to be improved.

RCM method can reduce the bandwidth of sparse matrices [29]. Therefore, the coefficient matrix of (5) and (10) can be reordered by using RCM, which helps to improve the solution speed. For a 19,000-node calculation, the coefficient matrix of the current continuity equation is reordered by using the RCM method. Figure 2 shows that non-zero elements in the coefficient matrix become more compact after RCM reordering.

Figure 2. Comparison of coefficient matrix before and after RCM reordering: (**a**) Unsorted; (**b**) RCM-Sorted.

The coefficient matrix in (6) and (10) can also be preprocessed by using BMA [31], which reorders the matrices to enhance diagonal dominance. Then (10) can be transformed as:

$$\mathbf{RPACY} = \mathbf{RPB} \tag{11}$$

where $\mathbf{R}$ is the permutation matrix, $\mathbf{P}$ is the row scaling matrix, $\mathbf{C}$ is the column scaling matrix, and $\mathbf{Y} = \mathbf{C}^{-1}\boldsymbol{\rho}$. After $\mathbf{Y}$ is solved, the space charges $\boldsymbol{\rho}$ can be obtained. BMA reduces the condition number of the coefficient matrix, resulting in improved convergence behavior.

The accelerated solution procedure is shown in Figure 3. An inner loop and an outer loop are included in the iteration. The inner loop adjusts the surface charge distribution on AC and DC lines to ensure that the surface electric field does not exceed the corona onset threshold. The outer loop performs time-domain iteration until the convergence condition is reached. The acceleration process is applied to handle the coefficient matrices in the inner loop.

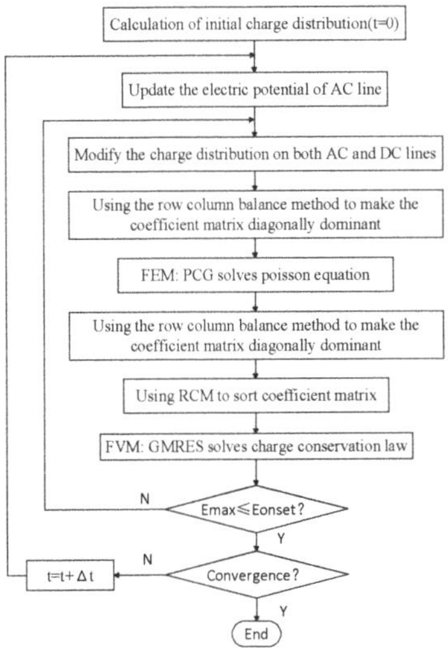

Figure 3. FEM-FVM accelerated solution procedure.

2.3. *Validity of Calculation Method*

For the model of ±800 kV HVDC and one-circuit 1000 kV HVAC transmission lines in the same corridor, there are 26,721 elements and 13,719 nodes in FEM mesh. The treatment of BMA can significantly reduce the condition numbers of the coefficient matrix. Table 1 shows the different computation times in a single iteration with a calculation tolerance of 1×10^{-17}. For the electric potential calculation, the condition number of FEM discretization equations decreases from 5.47×10^{18} to 3.22×10^{3}, and the computation time is decreased from 1.056 s to 0.173 s after BMA is used. For the space charges calculation, the condition number of the FVM discretization equations decreases from 2.63×10^{70} to 108.86, and the computation time is decreased from 0.578 s to 0.254 s after BMA and RCM are used. During the whole iterative process of ground-level electric field, there are approximately 20,000 outer iterations to achieve stable convergence. After acceleration, the total computation time is reduced from 32,680 s to 8,558 s, achieving an acceleration ratio of 3.8.

Table 1. Comparison of the accelerated solution for each iteration.

Calculation Method	Computation Time (s)	Condition Number
PCG (in FEM)	1.056	5.47×10^{18}
PCG with BMA (in FEM)	0.173	3.22×10^{3}
GMRES (in FVM)	0.578	2.63×10^{70}
GMRES with BMA and RCM (in FVM)	0.254	108.86

The actual transmission line model was also calculated using the calculation method. In the Baihetan–Zhejiang transmission lines, the structure of one section is ±800 kV and 500 kV HVDC lines on a single tower, and the AC circuit is a double-circuit vertical arrangement in reverse phase sequence. Table 2 compares the calculation results of the proposed method with those of the China Electric Power Research Institute (CEPRI) for the engineering model, which shows this method is also applicable to practical engineering [32].

Table 2. Comparison of Baizhe Engineering Line Results.

	Max Value of DC Component (kV/m)	Max Value (RMS) of AC Component (kV/m)
CEPRI	2.64	9.85
FEM-FVM with acceleration	2.90	9.71

3. Simulation Models and Design Limits

3.1. Simulation Models of Electric Field

The transmission lines considered in this paper include double-circuit 500 kV HVAC lines and one-circuit ±800 kV HVDC lines. HVAC lines can be arranged in an inverted triangular, vertical and horizontal configuration. The conductor parameters are listed in Table 3, where the sub-conductor types comply with the standard for "Round wire concentric lay overhead electrical stranded conductors" (GB/T 1179–2017 [33]). There are six kinds of phase sequence arrangements for double-circuit HVAC lines. The phase sequence of the left circuit is fixed as illustrated in Figures 4 and 5, and the phase sequences of the right circuit are listed in Table 4.

Table 3. Conductor Parameters.

Voltage	No. of Sub-Conductor	Sub-Conductor	Splitting Distance (mm)
±800 kV	6	JL/G3A-900/75	450
500 kV	4	LGJ-630/45	400

Table 4. Phase sequence arrangement.

Line Number	I	II	III	IV	V	VI
1/2/3	C/B/A	C/A/B	B/A/C	B/C/A	A/B/C	A/C/B

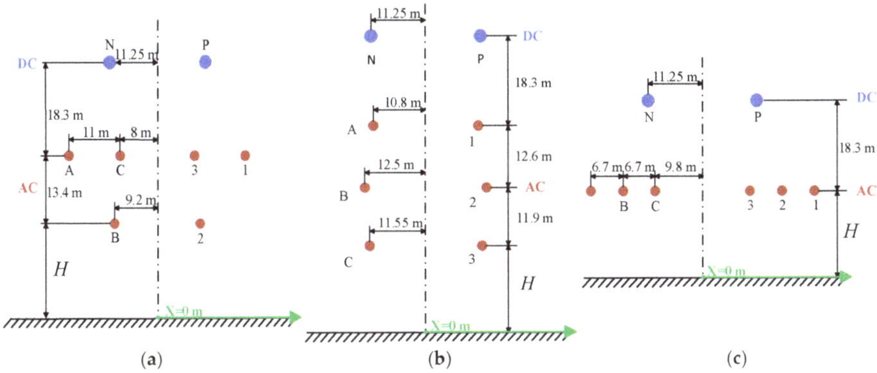

Figure 4. HVDC and HVAC lines on one tower: (**a**) Inverted Triangular Arrangement; (**b**) Vertical Arrangement; (**c**) Horizontal Arrangement.

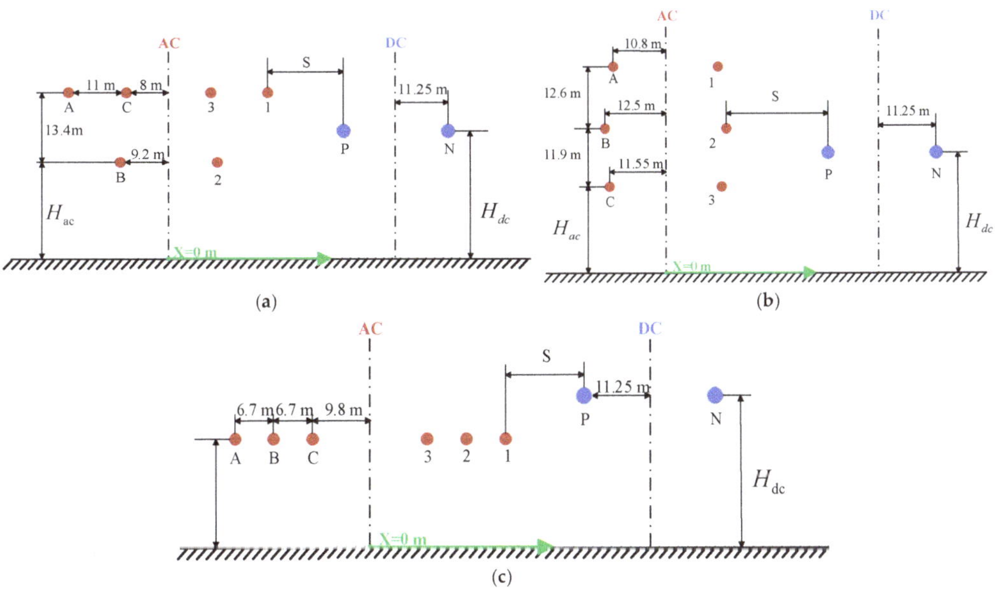

Figure 5. HVDC and HVAC lines on different towers: (**a**) Inverted Triangular Arrangement; (**b**) Vertical Arrangement; (**c**) Horizontal Arrangement.

For double-circuit 500 kV HVAC lines and one-circuit ±800 kV HVDC lines on different towers in the same corridor, illustrated in Figure 5, the minimum horizontal distance S between AC lines and DC lines is determined to be 20 m [34].

3.2. Design Limits

The limit values of the electric field, ion current density, audible noise (AN) and radio interference (RI) adopted in the paper in China are listed in Table 5, which are used to determine the conductor height and other geometry configuration for hybrid transmission lines in non-residential area or residential area [35]. The limit of RI or AN is at the position 20 m outside the positive pole for HVDC lines or outside the outer phase line for HVAC lines. RI is calculated using the CISPR method, and AN using the BPA method. In Table 5, weather conditions include dry and wet states, with corresponding surface roughness coefficients of 0.47 and 0.37, respectively. These values are used in the Peek equation to calculate the corona onset electric field of stranded wire. For AN, "sunny day" and "rain" refer to weather conditions. For RI, the 80%–80% rule refers to a statistical method ensuring that 80% of the appliances being tested meet the specified radio noise limit with at least 80% confidence [36].

Table 5. Electromagnetic environmental limits.

Factor	Situation	500 kV	±800 kV	
Electric field (kV/m)	Non-residential	10	dry	30
			wet	36
	Residential	7	dry	25
			wet	30
Ion current density (nA/m^2)	Non-Residential	-	100	
	Residential	-	80	
AN (dB(A))	sunny day	-	45	
	rain	55	-	
RI (dB(μV/m))	80%–80%	55	58	

In China, there is no national standard for the electromagnetic environment of HVAC-HVDC hybrid transmission lines, but the limits for separate HVAC lines and HVDC lines have been established. The normalized AC and DC electric field using their respective control values to determine the limit for hybrid electric fields is recommended [34]

$$C_H(x) = \frac{E_{DC}(x)}{E_{DCLim}} + \frac{E_{AC}(x)}{E_{ACLim}} \tag{12}$$

where E_{DCLim} and E_{ACLim} are the limits for DC and AC transmission lines, respectively, C_H denotes the normalized value of the hybrid electric field. The hybrid electric field under the hybrid lines must satisfy

$$C_H \leq 1 \tag{13}$$

Experimental studies on human perception and transient shock sensations under electric fields with varying proportions of AC and DC components indicate that the above method can be reliably used in engineering designs [10,23,37].

4. HVDC and HVDC Lines on One Tower

4.1. Main Design Constraint Factor

It is necessary to obtain the main design constraint that has the greatest impact on engineering design due to electromagnetic environment, in other words, to determine the most important factor among electric field, RI and AN during the design of HVAC-HVDC hybrid transmission lines. If the value of the factor is under the permissible limit, other electromagnetic factors will not be paid attention to in terms of the electromagnetic environment.

In order to obtain the main design constraint, the height H in Figure 4 is increased until only one factor is beyond the limit, which is the significant factor. Taking phase sequence I of double-circuit 500 kV AC lines as an example, the results of normalized hybrid electric field, ion current density, RI and AN for three different AC arrangements are shown in Table 6. Because the ion current density is very low due to the shielding effect of AC lines, it can be neglected during the design of hybrid lines on one tower. For the inverted triangular arrangement and vertical arrangement, the normalized hybrid electric field is more than 1; therefore, the electric field is the main constraint factor during the design. However, for the horizontal arrangement, the main design constraint factor is RI. More calculations show that the main factor influencing the design does not change with the variation of the phase sequence.

Table 6. The main design constraint factor for hybrid lines is on one tower.

Design Indicator			Triangular	Vertical	Horizontal
Increased Height (m)			10	10	14
C_H	Non-residential	Dry	1.23	1.16	0.65
		Wet	1.26	1.19	0.65
	Residential	Dry	1.75	1.16	0.90
		Wet	1.79	1.19	0.90
Ion current density (nA/m^2)		Dry	0.85	0.85	1.44
		Wet	1.02	0.98	1.82
RI @ DC side (dB(μV/m))			49.96	46.75	61.78
RI @ AC side (dB(μV/m))			47.59	46.2	55.92
Constraint factor			E Field	E Field	RI

4.2. Minimum Height Above the Ground

It is necessary to understand the minimum height for different structures in order to provide design references. By continuously increasing H and calculating the electromagnetic environment of the model, the minimum height is determined when all design indicators in Table 6 meet their respective limit requirements. Table 7 presents the minimum height corresponding to 6 kinds of phase sequences under inverted triangular, vertical and horizontal arrangements. For horizontal arrangement, the minimum height is excessively high due to RI. Therefore, horizontal arrangement is not recommended for HVAC-HVDC lines on one tower, and it is not included in the following discussion.

Table 7. Minimum height for three arrangements of hybrid lines on one tower.

Area Type	Arrangement	I	II	III	IV	V	VI
Non-residential	Triangular (m)	12.5	12	11	11	12	11
	Vertical (m)	12	12	12	12	12	12
	Horizontal (m)	32	31	31	28	32	27
Residential	Triangular (m)	16.5	15	14	14	16.5	14
	Vertical (m)	15	15	16	15	16	15
	Horizontal (m)	32	31	31	28	32	27

In non-residential areas, phase sequences III, IV and VI with a minimum height of 11 m are recommended for the inverted triangular arrangement. For the vertical arrangement, the impact of phase sequence on minimum ground clearance is negligible, and the minimum height of 12 m is suitable for all sequences.

In residential areas, phase sequences III, IV and VI with a minimum height of 14 m are also recommended for the inverted triangular arrangement. For vertical arrangement, phase sequences I, II, IV and VI are recommended with a minimum height of 15 m.

As an example, Figure 6 illustrates the ground-level electric field for two kinds of arrangements with phase sequence I in non-residential areas, where the zero X coordinate is the center of the AC transmission circuit. The hybrid electric field distribution of HVAC-HVDC lines on one tower is respectively illustrated as AC electric field, DC electric field and the superposition of AC and DC electric field. The results of the inverted triangular arrangement when H is 12.5 m are illustrated in Figure 6a. The results of the vertical arrangement when H is 12 m are illustrated in Figure 6b. For the inverted triangular and vertical arrangement, AC peaks are closer to the center, while DC peaks are far from the center. The staggered peak distribution is helpful to keep the normalized hybrid electric field below the limit. If the arrangement of conductors is the same, the peak values are higher under wet conditions than under dry conditions, because the corona onset electric field under wet conditions is lower.

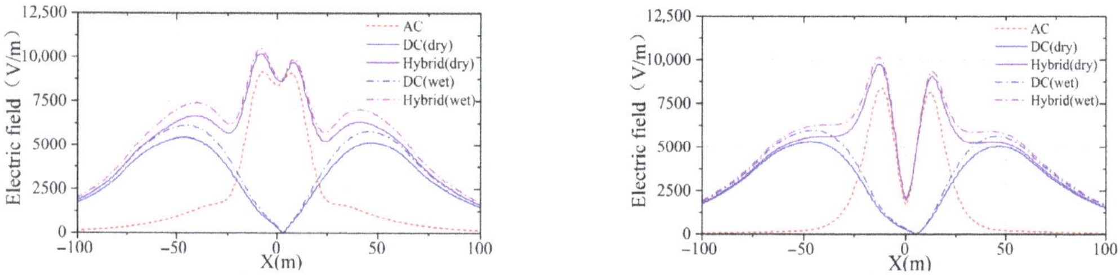

(**a**) (**b**)

Figure 6. Lateral distribution of ground-level hybrid electric field under DC-AC lines on one tower: (**a**) Inverted Triangular Arrangement; (**b**) Vertical Arrangement.

4.3. Corridor Width

Two key factors must be considered when determining the corridor width for hybrid transmission lines. The first one is the horizontal distance outside the edge conductor, which is 5 m for a 500 kV AC line side and 7 m for a ±800 kV DC line side [38,39]. The other one is the ground-level electric field. In China, the AC electric field should not be stronger than 4 kV/m, and the DC electric field should not be stronger than 15 kV/m at the edge of the corridor. Therefore, the corridor width is determined according to the normalized hybrid electric field

$$C_W = \frac{E_{AC}}{4} + \frac{E_{DC}}{15} \quad (14)$$

where at the edge of the corridor width

$$C_W = 1 \quad (15)$$

Figure 7 shows the hybrid electric field when the height H is 13 m. The corridor width is from the position x_3 to x_2 for the inverted triangular arrangement and from x_4 to x_1 for the vertical arrangement. In point of hybrid electric field, the corridor width of inverted triangular arrangement is smaller than that of vertical arrangement, but the corridor width gets larger based on the position of edge conductor, which is shown in Table 8. For six kinds of phase sequences, respectively, determined by the edge conductors and by normalized hybrid electric field, the phase sequence variation has a small effect on the corridor width. Finally, the inverted triangular arrangement is not recommended in reality, and the vertical arrangement is recommended, where the corridor width is determined by the normalized hybrid electric field C_W.

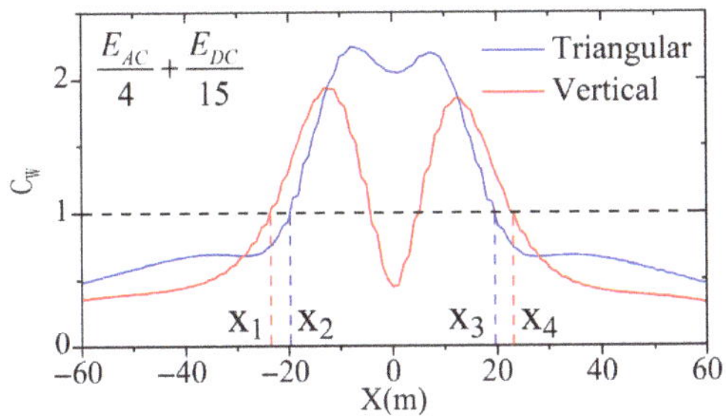

Figure 7. Normalized hybrid electric field for AC-DC lines on one tower with inverted triangular and vertical arrangements (sequence I).

Table 8. Corridor width for hybrid lines on one tower.

Phase Sequence	Triangular (m)		Vertical (m)	
	By C_W	By Edge Conductors	By C_W	By Edge Conductors
I	39	48	47	35
II	38			
III				
IV				
V				
VI	37			

5. Hvdc and Hvac Lines on Different Towers

5.1. Main Design Constraint

Table 9 lists the electric field, RI and AN for three kinds of arrangements in Figure 5, where the horizontal approaching distance S is 20 m, with the height of AC lines (*Hac*) 11 m and the height of DC lines (*Hdc*) 17 m. In non-residential areas, only the C_H value in Table 9 exceeds the limits in Table 5, indicating that the hybrid electric field is the main design constraint. In residential areas, both the ion current density under wet conductor conditions and C_H exceed the limits, so both the electric field and ion current density are the main design constraints. Calculation results for other phase sequences indicate that the constraint factor is not changed with phase sequence. Because the electric field is stronger under wet conditions than under dry conditions, the following discussion on the approaching distance, minimum height and corridor width will be under wet conditions.

Table 9. The main design constraint factor for AC-DC lines on different towers.

Design Indicator			Triangular	Vertical	Horizontal
C_H	Non-residential	Dry	1.13	1.09	1.09
		Wet	1.14	1.12	1.17
	Residential	Dry	1.59	1.51	1.44
		Wet	1.61	1.56	1.49

Table 9. *Cont.*

Design Indicator		Triangular	Vertical	Horizontal
Ion current density (nA/m^2)	Dry	43.05	35.64	36.74
	Wet	86.81	81.94	82.24
AN @ DC side (dB(A))	Dry	42.67	43.01	47.95
	Wet	42.97	41.17	41.58
AN @ AC side (dB(A))	Dry	41.88	42.59	48.75
	Wet	36.30	35.96	35.66
RI @ DC side (dB(μV/m))		54.36	50.68	51.02
RI @ AC side (dB(μV/m))		41.91	42.85	51.55
Constraint factor		E Field	E Field	E Field

5.2. Minimum Height and Corridor Width

For HVAC-HVDC lines on different towers, extensive calculations are required to determine the minimum height of AC and DC lines, represented by *Hac* and *Hdc*. The calculation procedure is shown in Figure 8. A set of calculations by fixing *Hac* (or *Hdc*) while varying *Hdc* (or *Hac*) are completed one by one.

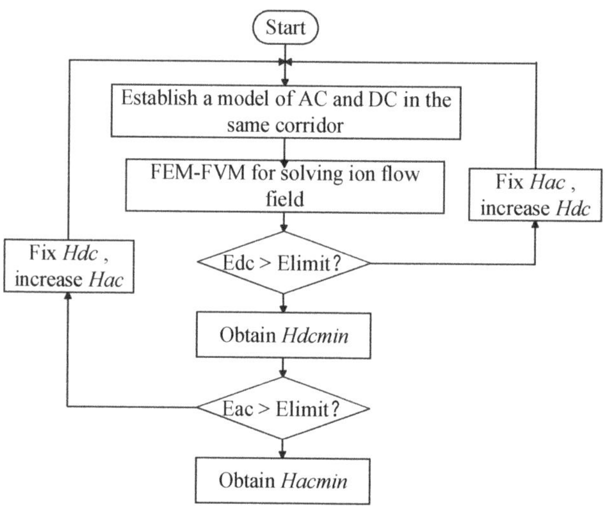

Figure 8. Flowchart to determine the minimum height of AC lines and DC lines.

For phase sequence I, the normalized hybrid electric fields with a wet conductor are shown in Figure 9. The maximum C_H in the AC area meets the limit requirements under the case of *Hac* 13 m and *Hdc* 16 m, and the maximum C_H in the DC area under the case of *Hac* 11 m and *Hdc* 19 m is at the threshold. Therefore, the minimum heights are respectively determined as *Hac* 13.5 m and *Hdc* 19.5 m with the maximum C_H 0.95.

The hybrid electric fields for six kinds of phase sequences under three arrangements are calculated. The minimum heights in non-residential and residential areas are presented in Table 10 with the corresponding corridor widths.

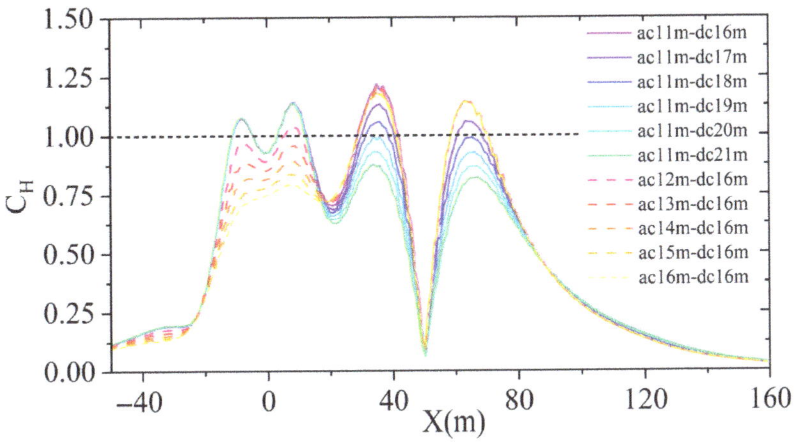

Figure 9. Normalized hybrid electric field for AC-DC lines on different towers with inverted triangular arrangement (phase sequence I).

In non-residential areas, phase sequence III for the inverted triangular arrangement is recommended, where the minimum height is 11.5 m for AC lines and 19 m for DC lines. The phase sequence VI for vertical arrangement is recommended, where the minimum height is 14 m for AC lines and 18 m for DC lines. The recommended phase sequences for horizontal arrangement are IV and VI, corresponding to the minimum height of 12 m for AC lines and 20 m for DC lines.

In residential areas, phase sequence III is recommended for the inverted triangular arrangement, where the minimum height is 15 m for AC lines and 22 m for DC lines. The phase sequence I for the vertical arrangement is recommended with a minimum height of 17 m for AC lines and 22.5 m for DC lines. The phase sequence I or V for the horizontal arrangement is recommended with a minimum height of 15 m for AC lines and 25 m for DC lines, or 16 m for AC lines and 24 m for DC lines.

For the inverted triangular and vertical arrangements, the corridor widths determined by edge conductors (93.5 m and 80.5 m, respectively) are smaller than those based on the hybrid electric field. Therefore, for these two arrangements, the corridor width should be determined by the hybrid electric field. However, for the horizontal arrangement, the corridor width determined by edge conductors is 101 m, which is larger than the results of phase sequences II, IV and VI. In order to discuss it in detail, the C_W of phase sequences IV and V is illustrated in Figure 10. The width for phase sequence V according to (16) is the distance between A and C, while for phase sequence IV, the width is the distance between B and C. It means that the electric field at the AC side of the phase sequence IV is very low, which is not a decisive factor for the corridor width. Therefore, the corridor width should be determined by the position of the outer AC line.

From the perspective of the minimum height, the inverted triangular arrangement is the best. In contrast, the vertical arrangement requires a higher height for AC lines, and the horizontal arrangement requires a higher height for DC lines.

From the perspective of the corridor width, the vertical arrangement of AC lines is more advantageous among the three arrangements, because the corridor width is the smallest. For the case including residential buildings, which is not considered in this paper, the electric field on the platform of the buildings where people can reach should also be calculated, because the new standard about DC electric field was published in 2020 [40].

Table 10. Minimum height and corridor width for AC-DC lines on different towers.

Arrangement	Area	Phase Sequence	*Hac* (m)	*Hdc* (m)	Corridor Width (m)
Triangular	Non-residential	I	13	19.5	108
		II	11.5	19.5	108
		III	11.5	19	107
		IV	12	19	108
		V	13	19	107
		VI	12	19	108
	Residential	I	18	23	105
		II	16.5	23	105
		III	15	22	105
		IV	16	22.5	104
		V	18	23	105
		VI	16	22.5	105
Vertical	Non-residential	I	14	19	104
		II	14	19.5	104
		III	15	19	104
		IV	14	19	104
		V	15	19	105
		VI	14	18	106
	Residential	I	17	22.5	101
		II	18	23	98
		III	19	23	101
		IV	19	24	99
		V	21	25	101
		VI	19	24	99
Horizontal	Non-residential	I	13	20	127
		II	13	21	127
		III	13	20	126
		IV	12	20	127
		V	12	20	127
		VI	12	21	127
	Residential	I	15	25	124
		II	17	25	70 (101)
		III	17	24	101
		IV	17	25	70 (101)
		V	16	24	121
		VI	17	24	99 (101)

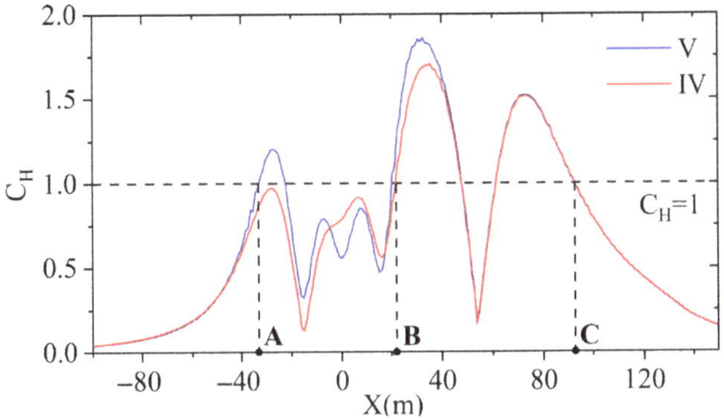

Figure 10. Normalized electric field for phase sequences IV and V with horizontal arrangement.

5.3. Impact of Approaching Distance

In order to analyze the influence of the approaching distance, the heights of phase sequence I in non-residential areas in Table 10 are adopted. The normalized hybrid electric fields are shown in Figure 11 for different approaching distances of 20 m, 30 m, 40 m, 50 m, 60 m and 70 m.

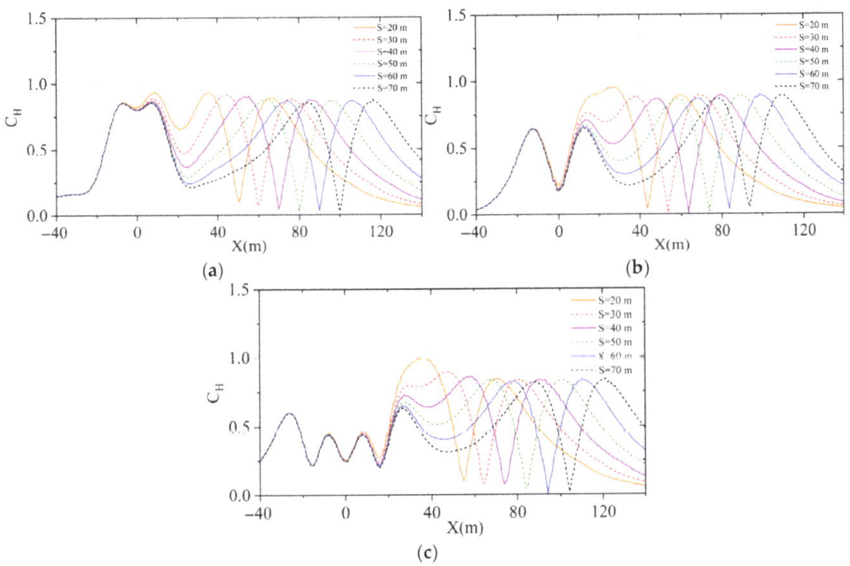

Figure 11. Normalized hybrid electric field for AC-DC lines (sequence I) with different approaching distance: (**a**) Inverted Triangular Arrangement; (**b**) Vertical Arrangement; (**c**) Horizontal Arrangement.

When the approaching distance is larger than 60 m, the normalized hybrid electric field is almost unchanged with the variation of S. The approaching distance is smaller, the more significant the coupling between AC lines and DC lines is. For vertical arrangement and horizontal arrangement, the maximum of the normalized electric field C_{Hmax} decreases significantly with the increase of approaching distance when S is smaller than 40 m. But for the inverted triangular arrangement, the maximum of the normalized electric field decreases slightly. Therefore, the minimum height should be related to the approaching distance between AC lines and DC lines.

6. Conclusions

In order to effectively analyze the electric field of hybrid HVAC and HVDC transmission lines in the same corridor, the FEM-FVM method is accelerated by the conjugate gradient method and generalized minimal residual method with pre-conditioning treatment. With the help of BAM and RCM, the computation speed can be increased by more than 3.8 times.

One-circuit ±800 kV DC lines and double-circuit 500 kV AC lines with inverted triangular, vertical and horizontal configurations under six kinds of different phase sequences are analyzed, and the calculation results, including hybrid electric field, ion current density, audible noise, and radio interference, are obtained. Based on the limit values in China, the main design constraint factor is determined. For HVAC-HVDC lines on one tower, the hybrid electric field is the most important factor for inverted triangular and vertical AC arrangements, but RI is the main constraint factor for horizontal arrangements. For HVAC-HVDC lines on different towers, the hybrid electric field in non-residential areas is the main factor influencing the design, and ion current density should be additionally considered in residential areas.

Different conductor arrangements can result in different results of minimum height and corridor width due to the distribution of hybrid electric fields. For HVAC-HVDC lines on one tower, the inverted triangular arrangement is the best choice to obtain lower height and smaller corridor width in the view of electric field, with recommended minimum heights of 11 m (phase sequence III, non-residential) and 14 m (phase sequence III, residential), and a corridor width of 37–39 m. For HVAC-HVDC lines on different towers, the phase sequence III inverted triangular configuration is better to obtain lower minimum height (Hac = 11.5 m and Hdc = 19 m for non-residential area, Hac = 15 m and Hdc = 22 m for residential area). The vertical arrangement is recommended to obtain a smaller corridor width of HVAC-HVDC lines in reality, and the horizontal arrangement is not recommended due to a wider corridor width. For vertical and horizontal configurations, increasing the approaching distance S within 30–40 m is recommended to effectively reduce the C_{Hmax}. For the inverted triangle configuration, increasing S has a limited impact on reducing C_{Hmax} and the minimum approaching distance of S = 20 m is recommended.

Author Contributions: Conceptualization and Writing—original draft, J.X.; Programming and Writing—original draft, C.H.; Methodology and Project administration, J.T.; Data curation and Formal analysis, H.W.; Supervision and Writing—review & editing, T.L. All authors have read and agreed to the published version of the manuscript.

Funding: This research was funded by PowerChina Sichuan Electric Power Engineering Corporation and The APC was funded by PowerChina Sichuan Electric Power Engineering Corporation.

Data Availability Statement: The original contributions presented in this study are included in the article. Further inquiries can be directed to the corresponding author(s).

Conflicts of Interest: Authors Jinyuan Xing, Jun Tian and Hao Wu are employed by PowerChina Sichuan Electric Power Engineering Corporation. The remaining authors declare that the research was conducted in the absence of any commercial or financial relationships that could be construed as a potential conflict of interest. The authors declare that this study received funding from PowerChina Sichuan Electric Power Engineering Corporation.

References

1. Sarma, M.P. *Corona Performance of High-Voltage Transmission Lines*; Crown Publication cc: Johannesburg, South Africa, 2011; pp. 12–13.
2. Larruskain, D.M.; Zamora, I.; Abarrategui, O.; Aginako, Z. Conversion of AC distribution lines into DC lines to upgrade transmission capacity. *Electr. Power Syst. Res.* **2011**, *81*, 1341–1348. [CrossRef]

3. El-Deen, S.K.; Abdel-Salam, M.; El-Mohandes, M.T. Analysis of corona on hybrid DC and AC overhead transmission lines. *IEEE Trans. Plasma Sci.* **2023**, *51*, 2926–2941. [CrossRef]
4. Hedtke, S.; Bleuler, P.; Franck, C.M. Outdoor investigation of the corona characteristics and audible noise of a hybrid AC/DC overhead line. *IEEE Trans. Power Del.* **2021**, *36*, 3309–3317. [CrossRef]
5. Fayazi, M.; Saffarian, A.; Joorabian, M.; Monadi, M. Analysis of induced components in hybrid HVAC/HVDC transmission lines on the same tower for various fault conditions. *Electr. Power Syst. Res.* **2024**, *226*, 109884. [CrossRef]
6. Fayazi, M.; Saffarian, A.; Joorabian, M.; Monadi, M. An AI-based fault detection and classification method for hybrid parallel HVAC/HVDC overhead transmission lines. *Electr. Power Syst. Res.* **2025**, *238*, 111083. [CrossRef]
7. Xia, J.; Ma, Z.; He, Z.; He, X.; Zhao, L.; Shen, B.; He, K.; Bian, X. Experimental and theoretical study of corona inception threshold in rod to plane electrodes under AC/DC combined voltage: Phase corresponding to AC voltage. *IEEE Trans. Dielectr. Electr. Insul.* **2023**, *30*, 403–412. [CrossRef]
8. Chartier, V.L.; Sarkinen, S.H.; Stearns, R.D.; Burns, A.L. Investigation of corona and field effects of AC/DC hybrid transmission lines. *IEEE Trans. Power App. Syst.* **1981**, *PAS-100*, 72–80. [CrossRef]
9. *IEEE Standard 2819-2022*; IEEE Recommended Practice for Measuring Method of Electromagnetic Environment for the Corridor of High-Voltage Overhead Power Transmission Lines in Parallel Mixed with Alternating Current and Direct Current. IEEE: New York, NY, USA, 2022.
10. Xie, L.; Lu, J.; Yang, Y.; Wu, G.; Ju, Y. Study on human perception experiments and design limits of the hybrid electric field of parallel HVAC and HVDC transmission lines. *Proc. CSEE* **2024**, *44*, 776–789.
11. Maruvada, P.S.; Drogi, S. Field and ion interactions of hybrid AC/DC transmission lines. *IEEE Trans. Power Del.* **1988**, *3*, 1165–1172. [CrossRef]
12. Clairmont, B.A.; Johnson, G.B.; Zaffanella, L.E.; Zelingher, S. The effect of HVAC-HVDC line separation in a hybrid corridor. *IEEE Trans. Power Del.* **1989**, *4*, 1338–1350. [CrossRef]
13. Abdel-Salam, M.; El-Mohandes, M.T.; El-Kishky, H. Electric field around parallel DC and multi-phase AC transmission lines. *IEEE Trans. Electr. Insul.* **1990**, *25*, 1415–1422. [CrossRef]
14. Zhao, T.; Sebo, S.A.; Kasten, D.G. Calculation of single phase AC and monopolar DC hybrid corona effects. *IEEE Trans. Power Del.* **1996**, *11*, 1454–1462. [CrossRef]
15. Yang, Y.; Lu, J.; Lei, Y. A calculation method for the hybrid electric field under UHVAC and UHVDC transmission lines in the same corridor. *IEEE Trans. Power Del.* **2010**, *25*, 1146–1153. [CrossRef]
16. Li, W.; Zhang, B.; He, J.; Zeng, R.; Li, X.; Wang, Q. Calculation of ion flow field of AC-DC hybrid transmission line. *IET Gener. Transm. Distrib.* **2009**, *3*, 911–918. [CrossRef]
17. Yin, H.; He, J.; Zhang, B.; Zeng, R. Finite volume-based approach for the hybrid ion-flow field of UHVAC and UHVDC transmission lines in parallel. *IEEE Trans. Power Del.* **2011**, *26*, 2809–2820. [CrossRef]
18. Zhou, X.; Cui, X.; Lu, T.; Zhen, Y.; Luo, Z. A time-efficient method for the simulation of ion flow field of the AC-DC hybrid transmission lines. *IEEE Trans. Magn.* **2012**, *48*, 731–734. [CrossRef]
19. Straumann, U.; Franck, C.M. Ion-flow field calculations of AC/DC hybrid transmission lines. *IEEE Trans. Power Del.* **2013**, *28*, 294–302. [CrossRef]
20. Guillod, T.; Pfeiffer, M.; Franck, C.M. Improved coupled ion-flow field calculation method for AC/DC hybrid overhead power lines. *IEEE Trans. Power Del.* **2014**, *29*, 2493–2501. [CrossRef]
21. Zhang, B.; Mo, J.; He, J.; Zhuang, C. A time-domain approach of ion flow field around AC–DC hybrid transmission lines based on method of characteristics. *IEEE Trans. Magn.* **2016**, *52*, 7205004. [CrossRef]
22. Qiao, J.; Zou, J.; Yuan, J.; Li, B. Electric Field Calculation of HVAC and HVDC Transmission Lines on the Same Tower With Domain Decomposition Method and High Order Element. *Power Syst. Technol.* **2017**, *41*, 335–341.
23. Li, B.; Qiao, J.; Zou, J.; Yuan, J. Analysis of Configuration and Corridor Width of HVAC and HVDC Transmission Lines on Same Tower. *Power Syst. Technol.* **2017**, *41*, 353–359.
24. Qiao, J.; Zhang, P.; Zhang, J.; Lu, Y.; Zou, J.; Yuan, J.; Huang, S. An iterative flux tracing method without Deutsch assumption for ion-flow field of AC/DC hybrid transmission lines. *IEEE Trans. Magn.* **2018**, *54*, 7200504. [CrossRef]
25. Ma, A.; Chen, J. Analysis of the three-dimensional hybrid electric fields and the related electrostatic induction effects of 800 kV DC and 500 kV AC parallel transmission lines considering environmental factors. *Electr. Power Syst. Res.* **2018**, *163*, 638–646. [CrossRef]
26. Tian, Y.; Huang, X.; Tian, W.; Cao, W.; Zhu, Y.; Zhao, L.; Zhang, Y. Time-domain mixed hybrid finite element method for analyzing ion-flow field of HVDC and HVAC transmission lines. *Electr. Mach. Control* **2019**, *23*, 85–94.
27. Xu, Q.; Liu, P.; Dinavahi, V. Parallel finite element computation of time-varying ionized field around hybrid AC/DC lines via fine-grained domain decomposition. *IEEE Access* **2020**, *8*, 91248–91256. [CrossRef]
28. Shen, N.; Su, Z.; Lu, T. The influence of charge characteristics of suspension droplets on the ion flow field in different temperatures and humidity. *Plasma Sci. Technol.* **2022**, *24*, 044004. [CrossRef]

29. Shen, N.; Su, Z.; Zou, Y.; Lu, T. Influence of humidity on the charge characteristics of suspension droplets and the characteristics of ion flow field. *Trans. China Electrotech. Soc.* **2022**, *37*, 3422–3430.
30. Ferziger, J.H.; Perić, M. *Computational Methods for Fluid Dynamics*, 3rd ed.; Springer: Berlin, Germany, 2002.
31. Duff, I.S.; Koster, J. On algorithms for permuting large entries to the diagonal of a sparse matrix. *SIAM J. Matrix Anal. Appl.* **2001**, *22*, 973–996. [CrossRef]
32. Xie, L. *Study on Electromagnetic Environment of Baihetan-Zhejiang ±800 kV Transmission Line with Other DC and AC Transmission Lines on the Same Tower or Parallel in the Adjacent Corridors*; Technical Report; China Electric Power Research Institute: Beijing, China, February 2023.
33. *GB/T 1179-2017*; Round Wire Concentric Lay Overhead Electrical Stranded Conductors. General Administration of Quality Supervision, Inspection and Quarantine of the People's Republic of China: Beijing, China, 2017.
34. Wu, G.; Yu, J.; Guo, X.; Ju, Y. Study on least approaching distances between ±800 kV DC transmission line and 1000 kV AC transmission line in a hybrid corridor. *Electr. Power* **2007**, *40*, 22–26.
35. *China DL/T 5582-2020*; Code for Electrical Design of Overhead Transmission Lines. China Planning Press: Beijing, China, 2020.
36. *CISPR Publication 18-2*; Radio Interference Characteristics of Overhead Power Lines and High-Voltage Equipment, Part 2: Methods of Measurement and Procedure for Determining Limits. IEC: Geneva, Switzerland, 2017.
37. *China T/CEC 688–2022*; Design and Control Standard for Hybrid Electric Field of EHV/UHV Transmission Lines. CEC: Beijing, China, 2022.
38. *China GB 50545*; Code for Design of 110 kV~750 kV Overhead Transmission Line. China Planning Press: Beijing, China, 2010.
39. *China GB 50790*; Code for Design of ±800 kV DC Overhead Transmission Line (Partial Revision). China Planning Press: Beijing, China, 2020.
40. *China GB 39220*; Limits and Monitoring Methods of Total Electric Field of DC Transmission Project. China Environmental Publishing Group Co., Ltd.: Beijing, China, 2020.

energies

MDPI

Article

Electrical–Thermal Aging Performance of PAH-Modified Interfacial Coating Agent for HVDC Cable Accessory

Wenbo Zhu [1,†], Kaulya Pathiraja [2,†], Xu Guo [2], Baojun Hui [1], Mingli Fu [1], Linjie Zhao [1], Yuhuai Wang [2] and Jin Li [2,*]

1 Electric Power Research Institute, China Southern Power Grid Company, Guangzhou 510080, China; zhuwb@csg.cn (W.Z.)

2 Key Laboratory of Smart Grid of Education Ministry, School of Electrical and Information Engineering, Tianjin University, Tianjin 300072, China; pathirajakaulya@tju.edu.cn (K.P.)

* Correspondence: lijin@tju.edu.cn

† These authors contributed equally to this work.

Abstract: A novel interfacial coating agent was developed by modifying silicone oil with polycyclic aromatic hydrocarbons (PAHs) to enhance the insulation performance of HVDC cable accessories. This study investigates the effects of corona and hot–cold cycle aging on the DC breakdown characteristics of the Cross-Linked Poly Ethylene and Ethylene Propylene Diene Monomer (XLPE/EPDM) interface. Interfacial breakdown tests, infrared spectroscopy, and a microstructural analysis were employed to investigate aging mechanisms. The results show that PAH-modified silicone oil significantly increases the breakdown voltage, with 2,4-dihydroxybenzophenone ($C_{13}H_{10}O_3$) identified as the optimal additive via quantum chemical calculations (QCCs). Even after aging, the modified interface maintains its superior performance, confirming the long-term reliability of the coating.

Keywords: HVDC cable accessory; interface breakdown; coating agent; corona aging; hot–cold thermal cycling; XLPE; EPDM

1. Introduction

China has abundant offshore wind energy, which is a key pillar for achieving the country's energy transition and advancing the "dual carbon" strategy. The optimal method for transmitting large-scale, reliable power from renewable sources is flexible high-voltage direct current (HVDC) transmission. Among these, the subsea cable terminal, which connects subsea cables to onshore equipment, is a critical component. Due to operational conditions and differences in insulation material performance, the terminal interface is prone to breakdowns, limiting large-scale transmission development. Improving the insulation of these HVDC cables directly increases the feasibility of new large-scale projects, as well as reducing the maintenance costs of existing projects. The main challenges faced by the existing technology can be summed down to the insulation layer of the cable terminal ends degrading with time. Factors such as extreme temperatures (both high and low) and extremely high pressures, which the cables are subjected underground, cause accelerated mechanical wear. Chemical interactions between the multiple layers of materials used to construct these cable terminals interact with each other, causing each layer to provide suboptimal insulation [1,2]. Although these effects are important, the most important factor is the constant high voltage (100 kV~800 kV) that these cables are handling. This causes a multitude of problems, such as space charge accumulation, which in turn causes partial discharges and electrical treeing. It is important to study the effect of high electric fields on these insulation layers in order to improve them [2–4].

In modern HVDC cable accessory insulation, a go-to solution is using Cross-Linked Poly Ethylene (XLPE) and Ethylene Propylene Diene Monomer (EPDM). These materials are favored for their excellent electrical and thermal resilience, although they do have their limits. Silicone oil has shown promising results when used as an insulation layer between an XLPE and EPDM interface [5,6]. Despite this, the performance of these systems eventually degrades with time, mainly due to long-term thermal and electrical stress [4,7–10].

The use of PAH small molecules as an additive to silicone oil to improve its electrothermal properties has also proven itself as an effective method. Studies have shown that doping with aromatic hydrocarbons at different mass ratios can increase the electrical breakdown properties of silicone oil at the oil-coated interfaces. Among these, 2,4-dihydroxybenzophenone has shown potential due to its molecular and thermal stability. However, the long-term behavior of such modified silicone oils under corona and hot–cold cycle aging has not been thoroughly examined. This is a considerable achievement when considering the scope of the industry and all possible applications. Since the existing research in this area is severely lacking, further testing on the limits of 2,4-dihydroxybenzopheone-doped silicone oil interfaces is necessary [11,12].

This study aims to evaluate the impact of corona discharge and hot and cold cycle aging on the microstructure, electrical breakdown characteristics, and molecular integrity of silicone oil-coated XLPE/EPDM interfaces. Both unmodified and 3% 2,4-dihydroxybenzophenone-doped silicone oils are investigated through DC breakdown testing, infrared spectroscopy, quantum chemical calculations, and microstructural changes at the oil-coated interface after both corona aging and hot–cold cyclic aging. This research gives insight into the effects of aging on the silicone-coated interface, the mechanisms involved, and the performance difference between standard silicone oil and silicone oil doped with 2,4-dihydroxybenzophenone.

1.1. Sample Preparation

1.1.1. XLPE Sample Preparation

The custom mold was cleaned with anhydrous ethanol and placed on PET film. The mold was covered with another layer of PET film. The plate vulcanizing machine was preheated to 110 °C for 5 min. The mold was slowly transferred into the machine, the upper plate was adjusted to lightly contact the mold (0 MPa pressure), and preheating was conducted for 2 min. After the particles were melted, the temperature was increased to 180 °C, and a pressure of 20 MPa was applied for 15 min. The mold was allowed to cool naturally under pressure to room temperature. The specimen was then removed and placed in a vacuum oven at 80 °C for 8 h to eliminate cross-linking byproducts. The final specimens were prepared as circular discs (300 μm thick, 100 mm diameter) for space charge measurements and square sheets (40 mm × 40 mm × 0.3 mm) for interface breakdown testing.

1.1.2. EPDM Sample Preparation

The roller distance was set to 1 mm and the temperature was set to 90 °C. EPDM particles were preheated in the roller mixer for 5 min until fully melted. DCP was gradually added to the mixer and blended for 5 min as a vulcanizing agent. The plate vulcanizer was preheated to 180 °C. The cleaned mold was placed between two PET films, the compound was added, and it was transferred to the vulcanizer. A pressure of 10 MPa was applied at 180 °C for 5 min. The mold was cooled under pressure to room temperature before demolding. The specimens were prepared as circular discs (300 μm thick, 100 mm diameter) and square sheets (40 mm × 40 mm × 0.3 mm). Figure 1 shows the appearance of the EPDM and XLPE samples.

Figure 1. (**a**) EPDM sample; (**b**) XLPE sample.

1.1.3. Preparation of Coating Modifying Agent

The coating agent (silicone oil) was modified via physical blending with a polycyclic aromatic hydrocarbon (PAH) compound. To ensure uniform dispersion and prevent agglomeration, 100 g of silicone oil was weighed in a beaker. The required amount of PAH, based on the desired mass percentage of the compound, was added, and the mixture was stirred magnetically for 40 min. The beaker was sealed with plastic wrap and sonicated for 10 min using an ultrasonic cleaner. The mixture was degassed in a vacuum chamber (10^{-4} Pa, 15 min) to remove air bubbles. The beaker was then sealed to prevent contamination. Figure 2 presents a simple flowchart of the coating agent modification process.

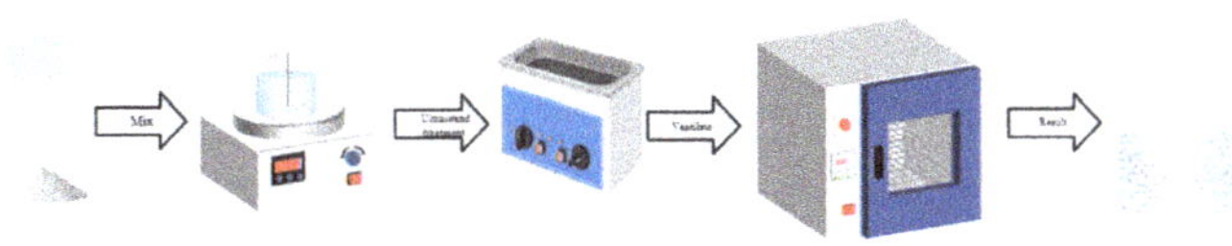

Figure 2. Flow chart of modified coating agent preparation.

The commercial XLPE particles used in this study were produced by Borealis (LS4201EHV), including a wire and cable extrusion-grade material with a density of 1.08 g/cm^3, a melt flow index of 20–30 g/min, and a molecular weight of 28.0532 g/mol. EPDM particles were sourced from Mitsui Chemical Co., Ltd. in Tokyo, Japan. (4045M), with a density of 1.85 g/cm^3 and a molecular weight of 190.32448 g/mol. The vulcanizing agent was dicumyl peroxide (DCP, purity 98%) produced by Nanjing Xinlian Chemical Co., Ltd. in Nanjing, China. The interface coating agent was dimethyl silicone oil ($[\text{-Si(CH}_3)_2\text{O-}]_n$, density: 0.963 g/mL) purchased from Shanghai Aladdin Reagent Co., Ltd. in Shanghai, China.

1.2. Experiment Procedures

1.2.1. Corona Aging Treatment

A needle and plate electrode pair was attached to the XLPE sample, and silicone oil was carefully applied over the XLPE sample carefully. The sample was placed on a horizontal surface and left undisturbed to settle evenly. To maintain consistent oil thickness, the initial mass of the XLPE sample (M_0) was measured using an electronic balance. After dripping the silicone oil, the mass (M_1) was recorded again. For a 0.3 mm thick sample, the mass difference ($M_1 - M_0$) was controlled to approximately 0.2 g. The mass difference ($M_1 - M_0 \approx 0.2$ g) was selected based on pre-calibration trials to achieve a consistent silicone oil layer thickness of approximately 0.3 mm across the 40 mm × 40 mm interface area. This mass-to-thickness mapping was determined assuming uniform density and surface distribution. Thickness uniformity was ensured by using a leveled deposition surface and visually inspecting for pooling or gaps. Although precise thickness measurements were not taken for each sample, samples with noticeable inconsistencies in surface coverage were discarded.

The needle electrode was used for corona discharge. The experimental apparatus can be seen in Figure 3. A maximum potential difference of 20 kV was applied to the two electrodes with a 5 mm gap in between the samples. The sample aging duration was 96 h. Ambient temperature during the aging process was 25 ± 2 °C, with relative humidity maintained at ~60%.

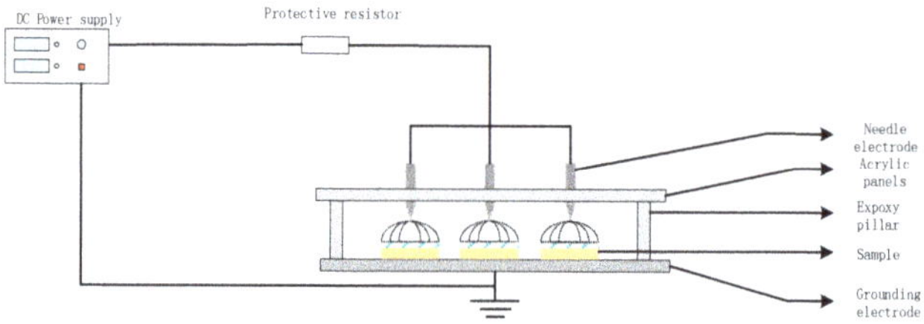

Figure 3. Corona aging system.

1.2.2. Hot and Cold Cycle Aging Treatment

The samples were placed in a custom-made high–low temperature test chamber shown in Figure 4. This is essential in order to maintain pressure on the interface while changing the temperature in a way that does not squeeze out the silicone oil in the interface. The temperature was alternated between a maximum of 90 °C and a minimum temperature of 0 °C, with each temperature interval maintained for 30 min. A total of 30 thermal cycles were performed. The goal of applying pressure during thermal cycling was to ensure consistent contact at the XLPE/EPDM interface while preventing excessive pressure that could displace the silicone oil layer. A constant pressure of approximately 0.05 MPa was maintained across all samples using the threaded screw mechanism shown in the apparatus, with real-time monitoring from a pressure sensor. Although the number of thermal cycles (30) was selected due to experimental time constraints, it is sufficient to represent a short-term accelerated aging condition and allows for meaningful comparative analysis between unmodified and modified silicone oils.

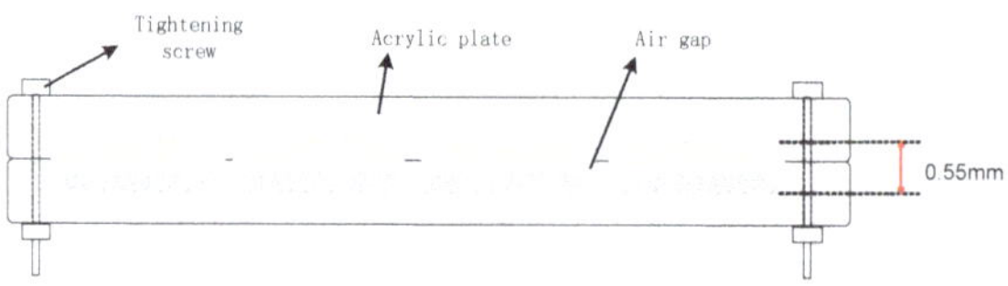

Figure 4. Sample fixing device.

1.2.3. FT-IR Analysis

Fourier Transform Infrared Spectroscopy (FTIR) is a technique that is used to analyze the chemical composition of materials. The method employs a Michelson interferometer to split light from a source, create an optical path difference, and generate interference patterns. The resulting data, encompassing all frequencies and intensities of the light source, undergoes a Fourier transformation via computer processing. This produces a frequency-dependent intensity spectrum. This enables us to identify the chemical composition by comparing the vibration frequency ranges by comparing it with known data [13,14].

To conduct this test, a Thermo Fisher Nicolet Nexus 470 FTIR spectrometer was used in attenuated total reflectance (ATR) mode. Measurement wavenumber range was set to 400–4000 cm^{-1}, with a resolution of 4 cm^{-1}. Since we were conducting the testing on liquid samples, a dry environment was maintained, a droplet of the sample was placed on a crystal surface, and the infrared spectra were obtained. The data was processed using the Origin 2019b software. To confirm the molecular structure of the selected interfacial coating agent, its IR spectrum is shown in Figure 5.

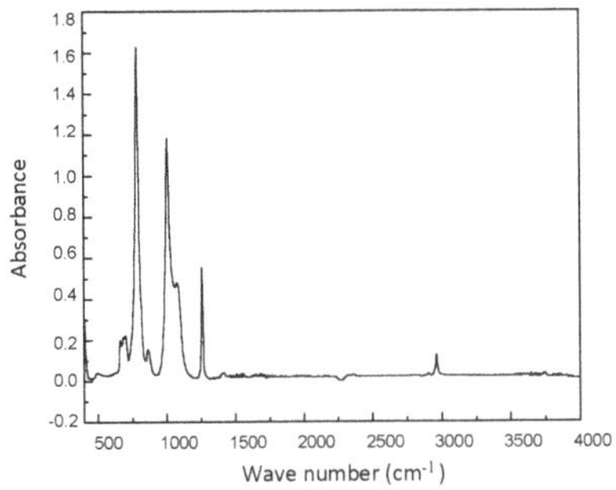

Figure 5. Infrared spectra of selected coating agents.

Dimethicones, also known as polydimethyloxanes, contain multiple silicone-oxygen (Si-O) units in their molecules as well as methyl (CH_3) clusters on the terminal and side groups. The infrared spectrogram of dimethicone oil has several significant vibrational absorption peaks, as shown in Table 1 [15]. By comparing the features of the IR spectra of the coating agent and dimethicone shown in Figure 5, it can be seen that the major vibrational absorption peaks of the selected coating agent coincide with those of dimethicone. This result verifies that the coating agent used is a high-purity dimethylsiloxane oil with no impurity components, further confirming the accuracy of its chemical structure and the singularity of its composition.

Table 1. Dimethyl silicone oil infrared absorption peaks and corresponding functional groups.

Assignment	Wave Number (cm^{-1})
$Si(CH_3)_3$	700
$Si(CH_3)_2$	780–840
C-H in $Si(CH_3)_3$	2960
$Si-CH_3$	1240–1280
Si-O-Si	1000–1130

1.2.4. DC Breakdown Testing

To closely simulate real-world HVDC submarine cable termination conditions (temperature: 10–90 °C, pressure: 0.05–0.3 MPa), two test setups were designed.

DC Breakdown Voltage Measurement: A needle–plate electrode configuration was used. Copper foil (50 μm thick) was cut and attached to the XLPE surface. The needle electrode (30° angle at the needle tip) and plate electrode were spaced 1 mm apart. Pressure and temperature were controlled via sensors and a thermal regulator. Voltage was applied incrementally at a ramp rate of 0.5 k V/s until breakdown occurred.

Pre-stressed DC Breakdown: A vertical DC pre-stress (15 kV/mm, 20 min) was applied to simulate charge accumulation. Post–pre-stress breakdown testing followed the same protocol. Figure 6 shows the experimental setup used for the DC breakdown voltage testing; on the top, we can also see a representation of the cross-sectional view of the EPDM/XLPE silicone oil-coated sandwich. A mechanical pressure press was used in conjunction with a pressure sensor to vary the pressure. A flat heating pad, as shown in the cross-sectional representation, was used to vary the temperature. Tests were repeated 10 times for each condition combination, and an average result was obtained. In all DC breakdown voltage measurements, the needle electrode was connected to the negative terminal of the DC power supply, and the plate electrode was grounded. Therefore, breakdown voltages are reported as negative values (e.g., −20.28 kV) to reflect the polarity of the applied electric field with respect to the ground electrode.

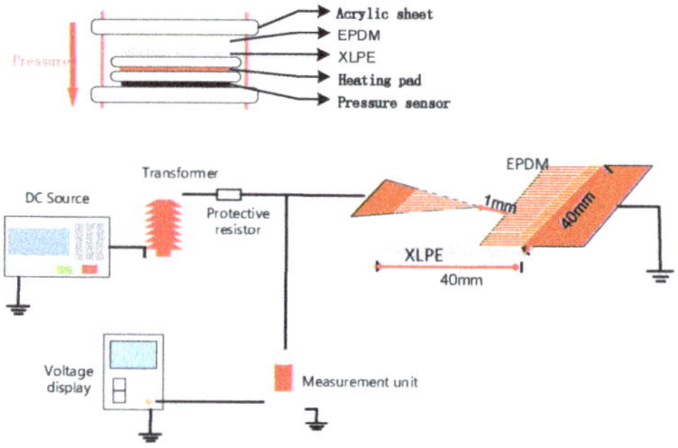

Figure 6. Device connection diagram for XLPE/EPDM interface breakdown testing.

1.2.5. Microstructure Analysis

A polarizing microscope was used to observe the microstructural changes in the samples before and after aging. The experimental setup is shown in Figure 7. The samples were placed on a transparent acrylic plate, illuminated with light, and observed through a computer-integrated S-EYE video system, which enabled us to record the findings.

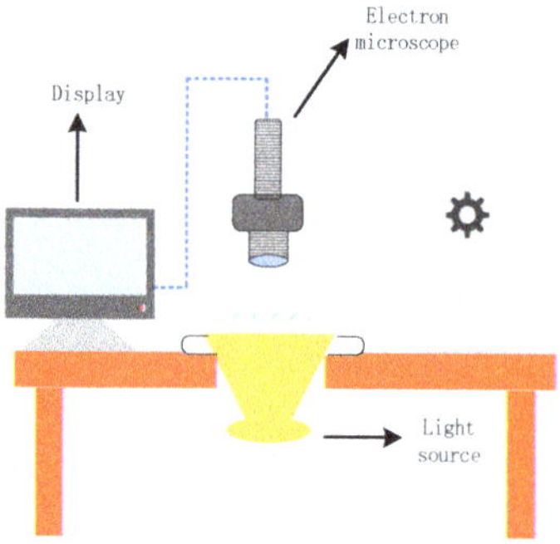

Figure 7. Polarized microscope observation system.

1.2.6. Quantum Chemical Computing

Quantum chemical calculations (QCCs) offer a powerful theoretical framework for analyzing the electronic structure and reactivity of molecules. In the context of dielectric materials, understanding how small molecular additives (such as PAHs) interact with high-energy charge carriers is essential for predicting their insulation performance under high-voltage stress. The ability of a molecule to trap high-energy electrons, as well as its ionization potential and electron affinity, are key indicators of its effectiveness in suppressing breakdown phenomena. Therefore, QCC was selected in this study to model and compare the frontier molecular orbitals (HOMO/LUMO levels) of candidate PAH modifiers and to assess the impact of aging-induced structural changes on their charge trapping capabilities.

With the significant increase in computational power, QCC has risen in popularity as a solution to the electronic structures of molecules, energy levels, and simulating properties of atoms and molecules. Phenomena such as polymer aging, breakdown, and electrical treeing are closely related to the movement and transfer of high-energy charge carriers. Quantum chemical calculations not only provide deep insights into the behavior of these carriers but also reveal their mechanistic roles in material performance evolution. Software such as VASP, Gaussian, and Q-Chem has proven itself as reliable software for QCC. In this experiment, we have used the Density Function Theory (DFT) proposed by Kohn and Hohenberg as the computational basis method for our platform [16,17].

In this simulation, we used Gaussian 09W as the computational platform and B3LYP functional with the 6-31G(d) basis set, in order to simulate the HOMO-LUMO energy levels of both aged and non-aged silicone oil. Analyzing the energy levels would give us a clear view of the underlying phenomenon.

2. Results

2.1. Effect of Corona Aging on Insulation Performance

Microstructure Observation

The sample was subjected to corona aging by exposing it to an electric field that was strong enough to ionize the surrounding air. A microscope was used to observe the surface of the specimens before and after corona aging, as shown in Figure 8. The results indicate that after being coated with silicone oil, the surface of the XLPE sample was smooth with no visible mechanical scratches or impurities. However, after 96 h of corona aging, the silicone oil layer dried out, and a few fine cracks appeared on the surface. Additionally, several circular pits were observed in the needle electrode region. This can be attributed to the high temperature and ozone generated during the corona discharge process.

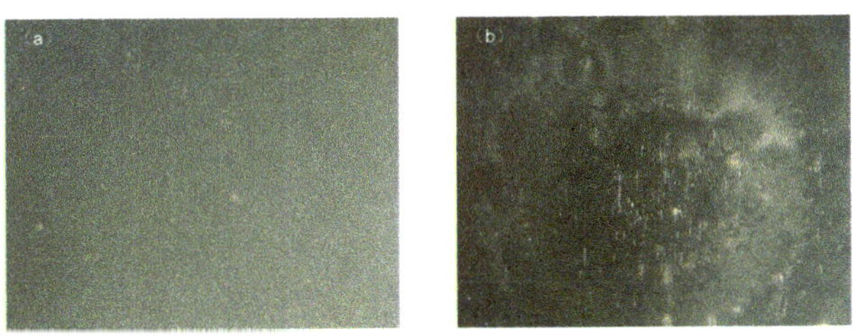

Figure 8. Surface microscopic images of specimens coated with unmodified silicone oil (**a**) before corona aging and (**b**) after corona aging.

Before corona aging, the surface of the modified silicone oil-coated specimens exhibited uniformly distributed 2,4-dihydroxybenzophenone crystals, which were embedded in the silicone oil matrix, forming a relatively stable structure. However, after corona aging, circular pits also appeared on the specimen's surface, and some doped crystals precipitated. This can be observed in Figure 9.

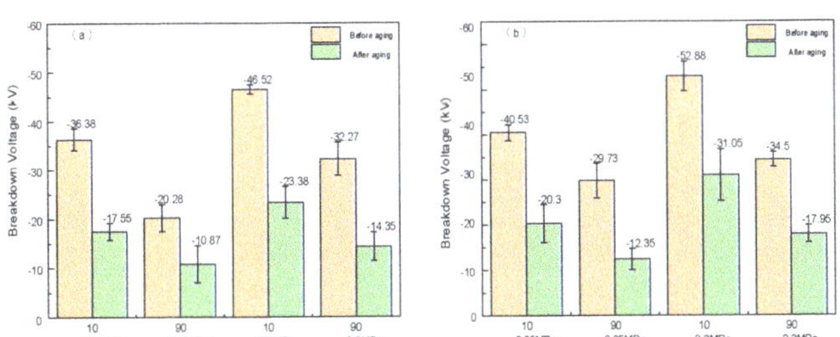

Figure 9. Microscopic images of the surface of specimens coated with modified silicone oil (**a**) before corona aging (**b**) after corona aging.

2.2. DC Breakdown Characteristics

The DC breakdown voltage after being subjected to corona aging was tested. This showed that the trends with varying temperature and pressure are the same as before being subjected to corona aging. The breakdown voltage decreases with increasing temperatures and increases with increasing pressures. The DC breakdown voltages are given in Figure 10a,b.

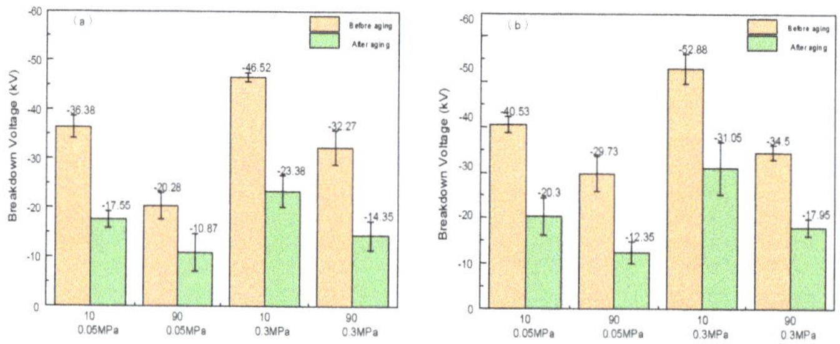

Figure 10. (**a**) Breakdown voltage after corona aging of XLPE/EPDM interface coated with unmodified silicone oil and (**b**) breakdown voltage after corona aging of XLPE/EPDM interface coated with 3% doped modified silicone oil.

It was observed that after subjecting the samples to corona aging, the breakdown voltage generally decreased. With the unmodified silicone oil under the conditions of 90 °C and 0.05 MPa, the breakdown voltage dropped from −20.28 kV to −9.87 kV, representing a 51.55% reduction. The modified silicone oil also displayed similar behaviors, where its breakdown voltage decreased from −29.73 kV to −12.35 kV, representing a 58.46% reduction.

2.3. Aging Mechanism Analysis

Following this, an infrared spectroscopy analysis was conducted with the corona-aged samples. Figure 11 shows the infrared spectra of unmodified silicone oil before and after corona aging.

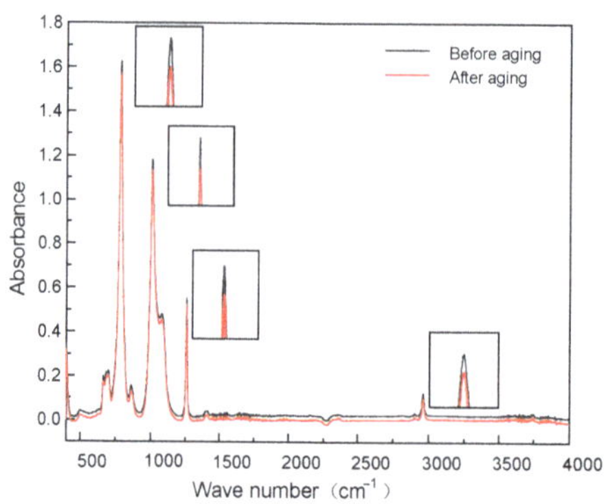

Figure 11. Infrared spectra of unmodified silicone oil before and after corona aging.

After comparing the values in Figure 11 with Table 1, along with the dimethyl silicone oil infrared absorption peaks and corresponding functional groups presented in Section 1.2.3 on FT-IR analysis, we can identify the functional groups present in the unmodified silicone oil. By comparing the two spectra "before aging" and "after aging" in Figure 11, we can observe whether any functional groups have been oxidized into other functional groups. The peaks at 2962.29 cm^{-1} and 2962.61 cm^{-1} correspond to the absorption vibration peaks of the saturated C-H bonds in -CH_3, while 1257.40 cm^1 and 1257.61 cm^{-1} correspond to the vibration absorption peaks of the Si-CH_3 bond. The peaks at 1007.74 cm^{-1} and 1007.48 cm^{-1} correspond to the vibration absorption peaks of the Si-O bond in Si-O-Si. The peaks at 784.97 cm^{-1} and 784.79 cm^{-1} correspond to the vibration absorption peaks of the Si$(CH_3)_2$ group. The peaks at 700.45 cm^{-1} and 696.07 cm^{-1} correspond to the vibration absorption peaks of the Si$(CH_3)_3$ group. After the process of corona aging, although no new peaks appeared in the spectra, the intensity levels of the major absorption peaks were significantly decreased.

The vibration absorption peak of C-H bonds decreased from 0.155 to 0.095. The vibration absorption peak of Si-CH_3 groups decreased from 0.552 to 0.451. The vibration absorption peak of Si-O bonds decreased from 1.182 to 1.106. The vibration absorption peak of Si$(CH_3)_2$ groups decreased from 1.625 to 1.514. The vibration absorption peak of Si$(CH_3)_3$ groups decreased from 0.223 to 0.169.

These results indicate that during the corona aging process, the aforementioned molecular chains and their functional groups and bond strengths have undergone a significant amount of degradation. However, no additional vibration absorption peaks were observed after the corona aging process, which also shows that the process has not formed or introduced new functional groups. Such degradation will have a notable impact on the insulation performance and cause a decrease in the breakdown voltage.

Figure 12 shows the infrared spectra of silicone oil doped with 3% 2,4-dihydroxybenzophenone by weight. A similar comparison can be done here with Table 1. The aforementioned molecular chains and their functional groups, and their bond strengths, have undergone a significant amount of degradation. From the figure, no new vibrational absorption peaks were observed after the corona aging process, which also indicates that no new functional groups were formed or introduced during the process. However, a degradation reaction was produced, resulting in some degree of structural changes in both the silicone oil molecule and the 2,4-dihydroxybenzophenone small molecule, which may lead to a decrease in the breakdown voltage. However, it can be observed that the IR spectra showed new vibrational absorption peaks in the range of 3100–3350 cm^{-1}, indicating the introduction of hydroxyl (-OH) groups. The peak of this group showed a different trend from the other peaks after corona aging. This is due to the aging of the -H functional groups on the benzene ring, which have been oxidized to hydroxyl (-OH) groups, causing an incremental change in absorbance from 0.320 to 0.711. No new vibrational groups were detected here, which confirms that no new functional groups have formed during aging. Therefore, corona aging leads to a decrease in the interfacial breakdown voltage of coated modified silicone oils in two ways:

1. The molecular chains of modified silicone oil undergo decomposition due to the impact of electrons from the needle electrode during corona aging.
2. New hydroxyl groups that form on the benzene rings are altering the chemical properties of the composition.

In order to further investigate whether the newly formed hydroxyl group (-OH) would have an effect on the properties of modified silicone oils, we analyzed 2,4-dihydroxybenzophenone before and after corona aging using quantum chemical calculations. The calculation results are shown in Figure 13.

According to Figure 13, which shows the electron energy level distribution of the $C_{13}H_{10}O_3$ molecules before and after aging, the HOMO and LUMO levels of the aged $C_{13}H_{10}O_3$ molecule are, respectively, −6.04 eV and −1.05 eV. The aged $C_{13}H_{10}O_3$ small molecule has an electron affinity energy of 1.05 eV and an ionization energy of 6.04 eV, which gives it a lower electron affinity energy and a wider energy gap compared to the original molecule. These results suggest a weakened ability of the aged molecule to trap high-energy electrons; therefore, when electrons enter the interface, the result is an increased probability of collisions between the high-energy electrons and the silicone oil molecule. At the same time, the increase in the energy gap means that the $C_{13}H_{10}O_3$ molecules in the excited state are difficult to excite, which means that the energy absorbed by the high-energy electrons cannot be safely released, and the captured high-energy electrons cannot be recycled again. The available $C_{13}H_{10}O_3$ molecules are reduced and their efficiency is lowered, and the high-energy electrons remaining increase, which makes it easier for them to collide with the silicone oil molecules. This aggravates the damage to the chain of the silicone oil molecules, and ultimately leads to the occurrence of breakdowns. It is worth noting that, although the LUMO energy level of the $C_{13}H_{10}O_3$ molecules increases and the HOMO energy level decreases after corona aging, the interfacial breakdown voltage is still higher than that of the unmodified silicone oil-coated interface, and there is still the ability to improve the interfacial breakdown voltage.

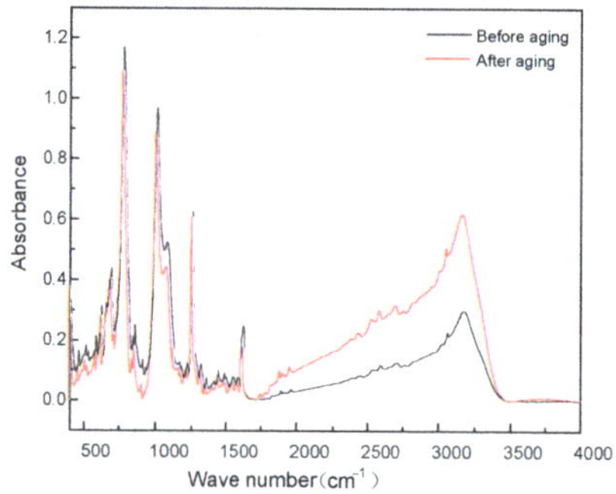

Figure 12. Infrared spectra of 3% modified silicone oil before and after corona aging.

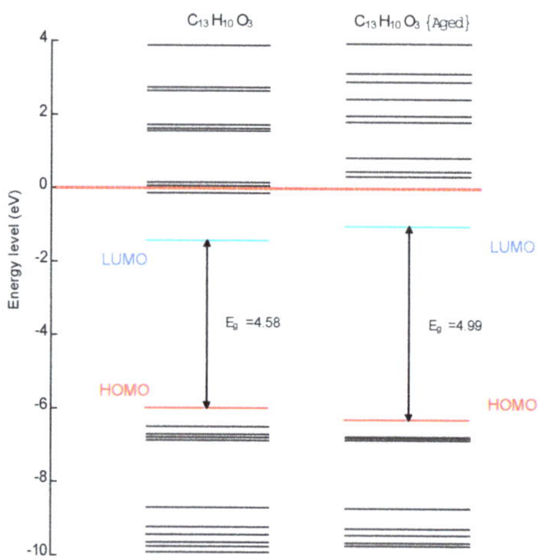

Figure 13. Energy level distribution of $C_{13}H_{10}O_3$ molecules before and after corona aging.

2.4. Effect of Thermal Cycling Aging on Insulation Performance

Microstructure Observation

Figure 14, presented below, shows the microscopic image of the interface of the specimens coated with unmodified silicone oil, both before and after hot–cold cycles. When the samples were subjected to hot–cold cycle aging, in the vicinity of the needle electrode, it was evident that while the EPDM/XLPE interface coated with undoped silicone oil pre-hot–cold cycle aging contained none to very few air bubbles, impurities could be seen. When it came to the aged samples, there was a considerable amount of bubble formation between the needle and the plate electrode.

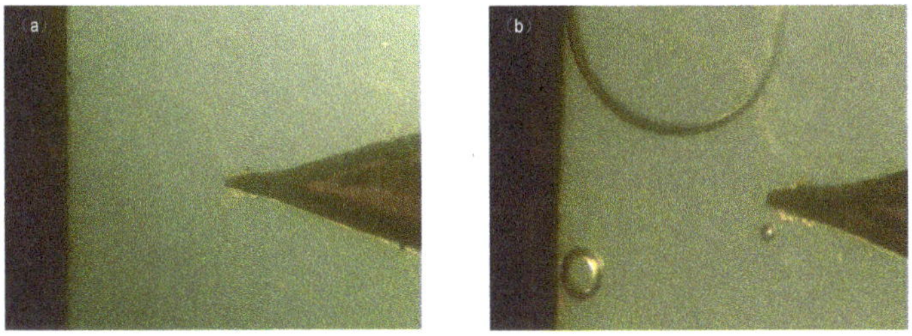

Figure 14. Surface microscopic images of specimens coated with unmodified silicone oil (**a**) before aging in hot and cold cycles and (**b**) after aging in hot and cold cycles.

Figure 15, presented below, shows the microscopic image of the interface of the specimens coated with modified silicone oil (3% W/W 2,4-dihydroxybenzophenone), both before and after the hot–cold aging cycles. We could observe uniformly distributed 2,4-dihydroxybenzophenone crystals and minimal to very little impurities and air bubbles. After being subjected to hot–cold cycle aging, we observed a considerable amount of air bubble formation, where we observed bubbles of varying sizes, while the number of embedded 2,4-dihydroxybenzophenone crystals significantly decreased.

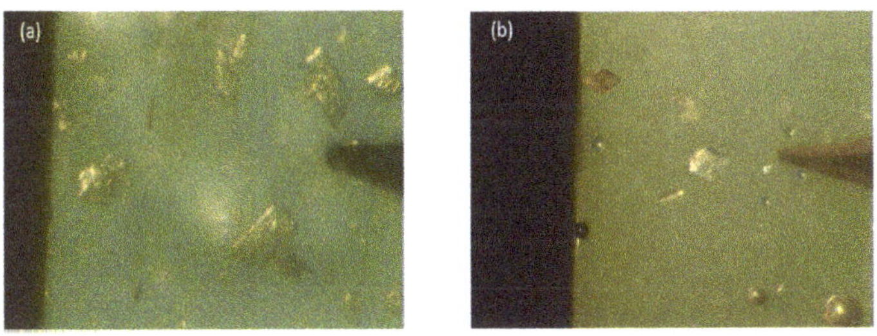

Figure 15. Microcosmic images of surface of specimens coated with modified silicone oil (**a**) before aging in hot and cold cycles and (**b**) after aging in hot and cold cycles.

2.5. DC Breakdown Characteristics

The results of DC breakdown testing of cold and hot cycle aged samples coated with modified and unmodified silicone oil interfaces are shown in Figure 16a,b, respectively. Despite undergoing hot and cold cyclic aging, the trends of the interfacial breakdown voltages of unmodified and modified silicone oil coatings under different temperature and pressure conditions are consistent, with both being inversely proportional to the temperature and positively proportional to the pressure. However, after hot and cold cyclic aging, the breakdown voltages of both the modified and unmodified silicone oil-coated samples decreased. It was also observed that under all the conditions, the silicone oil modified with 3% 2,4-dihydroxybenzophenone displayed the highest breakdown voltage readings.

These experimental results indicate that hot and cold cyclic aging significantly hinders the breakdown characteristics of silicone oil-coated interfaces. However, modifying the interface with 3% 2,4-dihydroxybenzophenone showed that it can mitigate this effect. It is noteworthy to mention that the breakdown voltage changes are more significant at lower interfacial pressures, while being less significant at higher interfacial pressures. This phenomenon is closely related to the more complete gas expulsion from the interface at higher pressures.

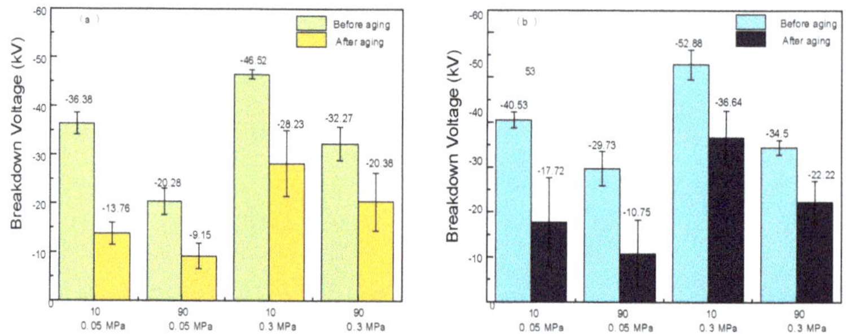

Figure 16. (**a**) Breakdown voltage of XLPE/EPDM interface coated with unmodified silicone oil after aging in hot and cold cycles; (**b**) breakdown voltage of XLPE/EPDM interface coated with 3% modified silicone oil after aging in hot and cold cycles.

2.6. *Aging Mechanism Analysis*

To further understand the effects of hot and cold cyclic aging on the silicone oil-coated XLPE/EPDM interface, infrared spectroscopy was used to conduct a deep dive into the structural changes that contributed to the changes in the breakdown voltage. Infrared spectroscopy was performed on both unmodified and modified 3% 2,4-dihydroxybenzophenone, each for 30 cycles of heat and cold. The results can be seen in Figures 17 and 18.

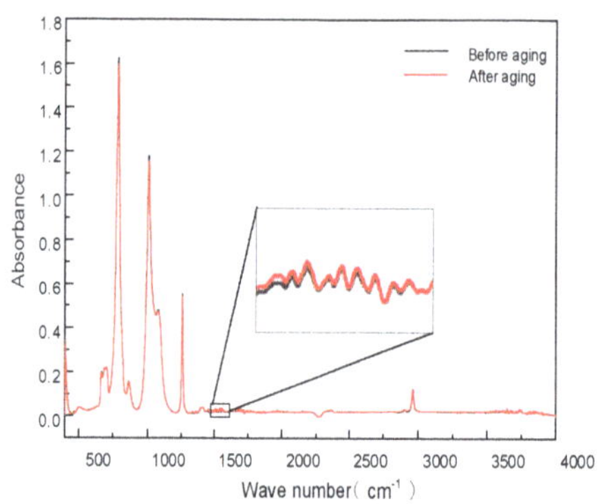

Figure 17. Infrared spectra of unmodified silicone oils before and after aging in hot and cold cycles.

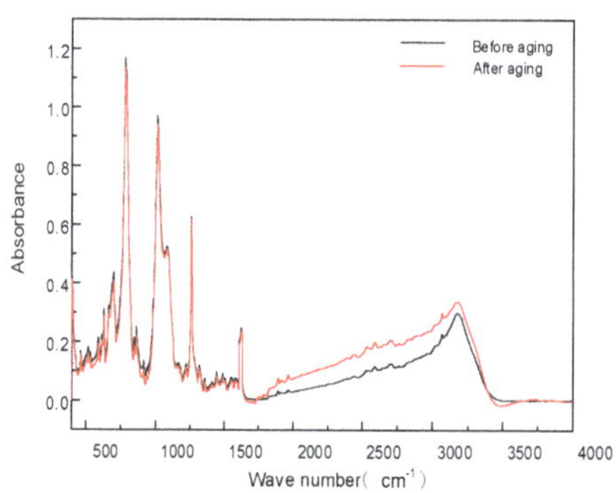

Figure 18. Infrared spectra of 3% modified silicone oil before and after aging in hot and cold cycles.

Observing Figure 17, which is presented above, showing the infrared spectra of the unmodified silicone oil before and after being subjected to hot–cold cycle aging, it was observed that the main vibrational absorption peaks exhibit slight changes after the process. The Si-O-Si vibrational absorption peak (1000–1100 cm^{-1}) decreased slightly from 1.181 to 1.165. The $Si(CH_3)_2$ vibrational absorption peak (780–840 cm^{-1}) decreased from 1.624 to 1.6. The vibrational peaks of C-H (around 2960 cm^{-1}), $Si\text{-}CH_3$ (1240–1280 cm^{-1}), and $Si(CH_3)_3$ (around 700 cm^{-1}) bonds remained unchanged. No new vibrational absorption peaks were observed. This indicates that the hot and cold cyclic aging had minimal effects on the degradation of the silicone-oil molecular chains.

Figure 18 contains the infrared spectra of the silicone oil modified with 3% 2,4-dihydroxybenzophenone. It can be observed that, unlike in the previous spectra in Figure 18, we can see a considerable change after the hot and cold cyclic aging process. The intensity of the main vibrational peak has decreased, which indicates molecular degradation of both the silicone oil and 2,4-dihydroxybenzophenone particles. Since no new absorption vibrational peaks were found, no new molecular structures were generated. The hydroxyl (-OH) vibrational absorption peak (3000–3300 cm^{-1}) increased from 0.291 to 0.322, indicating a slight increase in the hydroxyl group content. Although it is known from the above analysis of the corona aging mechanism that the increase in hydroxyl groups on the benzene ring may affect its quantum chemical properties and thus lead to changes in the properties of the modified silicone oils, we found that the slight increase in hydroxyl groups does not have a significant impact on the properties under hot and cold cyclic aging conditions. Despite a slight degradation, we conclude that this has a negligible influence on the DC breakdown at the interface. Thus, this is not the primary cause of the decrease in the breakdown voltage.

3. Discussion

Mechanisms of Bubble Formation and Interface Breakdown Development

From the microscopic morphology analysis of the XLPE/EPDM composite silicone-coated interface, a significant amount of bubble formation was observed after the hot and cold cyclic aging process. This phenomenon is attributed to the thermal expansion and contraction of the XLPE and EPDM. As the temperature changes, the contact between the two materials weakens, creating voids in between the two materials at the interface, and the process of silicone oil filling the interface is not perfectly efficient. The infrared spectra analysis showed that in both the modified and unmodified silicone oils, no new substances were formed, and only minor degradation was displayed. Therefore, it is inferred that the main reason for the decrease in the interfacial breakdown voltage is the formation of voids rather than chemical degradation.

The localized cavity discharge damage process and degradation mechanism analysis at the XLPE/EPDM interface are shown in Figure 19. As the breakdown field strength of air is much smaller than that of liquid and solid dielectrics, when bubbles appear in the XLPE/EPDM composite interface, the bubbles change the electric field distribution at the interface, making the electric field strength around the bubbles higher; at this time, the bubbles around the tip electrode begin to deform and move, as shown in the first stage of Figure 19. With the increase in voltage, the gas inside the bubble begins to ionize, and the partial discharge phenomenon initially occurs. Due to the self-recovering nature of air, the whole discharge briefly subsides after the partial discharge begins to occur at the bubble, as shown in the second stage of Figure 19. With the development of the discharge phenomenon and the continuous accumulation of electrons and ions within the bubble, the electric field within the bubble gradually loses stability until the bubble is filled with a large number of charged particles, and it loses of its insulating properties; at the same time, because of the concentration of the electric field near the bubble, the development of the discharge is intense, and the local temperature continues to rise, resulting in an accelerated aging process for the silicone oil and the two materials on both sides. The material's dielectric properties are gradually reduced, and the high temperatures further promote increases and expansion of the interface cavities, allowing for electronic kinetic energy to accumulate more easily, and finally, the interface breakdown phenomenon occurs as shown in Figure 19c.

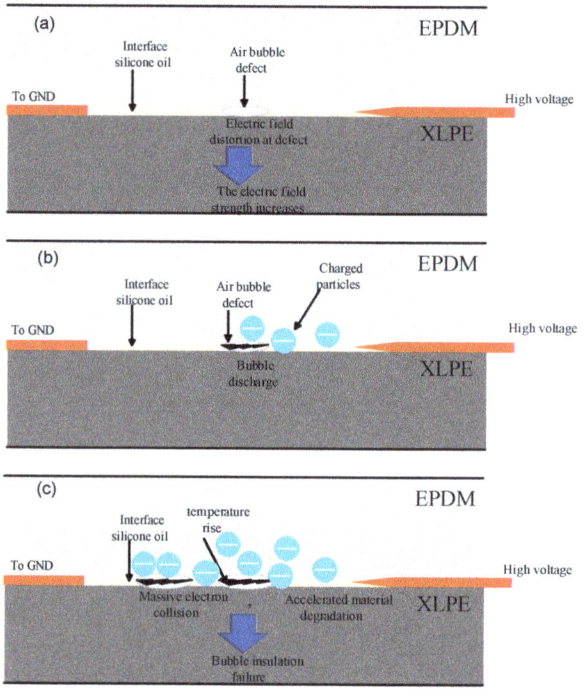

Figure 19. XLPE/EPDM interface bubble defect damage process (**a**) Stage I; (**b**) Stage II; (**c**) Stage III.

While this study investigated corona aging and hot–cold cycling independently to isolate their respective mechanisms, it is important to acknowledge that in real-world HVDC cable applications, these stress factors often occur concurrently. In such environments, the presence of thermal cycling-induced voids may exacerbate corona-induced degradation by providing localized weak points that concentrate electric fields and vice versa. In this study, the experiments were performed and the mechanisms of breakdown were studied in isolation for simplification purposes. A synergistic effect between these conditions is very likely, where these effects accelerate each other's impact on insulation breakdown. A great direction for future studies would be to investigate these synergistic effects.

4. Conclusions

In this study, we systematically investigated the mechanism of XLPE/EPDM interface property changes due to corona aging and thermal cycling aging. Emphasis was placed on confirming the excellent insulating properties of 2,4-dihydroxybenzophenone-doped modified silicone oil under aging conditions. Quantum chemical calculations revealed that the mechanism involves 2,4-dihydroxybenzophenone molecule trapping high-energy electrons and thus inhibiting the interfacial breakdown.

Corona Aging: Both unmodified and modified silicone oils exhibited a reduced breakdown voltage (51.55% and 58.46%, respectively) due to molecular chain degradation and the oxidation of benzene rings. FTIR revealed decreased bond strengths in the Si–O, Si–CH_3, and C–H groups, while quantum chemical calculations showed a decrease in the electron affinity energy and an increase in the energy gap in the aged $C_{13}H_{10}O_3$, resulting in a decrease in the ability of the molecule to trap high-energy electrons.

Hot and Cold Cyclic Aging: Breakdown voltage degradation was primarily attributed to mechanical void formation at the interface rather than chemical degradation. Hot–cold thermal expansion–contraction cycles formed large air bubbles, leading to partial discharges and localized electric field distortion. Hot and cold cyclic aging emphasized the critical role of physical defects (e.g., voids) over chemical changes, highlighting the need for improved interface quality and new bubble suppression strategies. Despite aging-induced performance declines, the modified silicone oil maintained superior breakdown characteristics compared to the unmodified oil, validating its potential for enhancing HVDC cable terminal reliability under the existing harsh operational conditions. These findings confirm that 2,4-dihydroxybenzophenone-modified silicone oil not only enhances the initial interface insulation performance but also significantly mitigates aging-related degradation, making it a promising candidate for long-term reliability in HVDC cable accessories.

Author Contributions: Conceptualization, W.Z.; validation, J.L.; formal analysis, J.L.; investigation, X.G.; resources, X.G. and W.Z.; data curation, Y.W.; writing—original draft preparation, K.P.; writing—review and editing, J.L. and X.G.; visualization, B.H.; supervision, M.F.; project administration, L.Z.; funding acquisition, J.L. All authors have read and agreed to the published version of the manuscript.

Funding: This research was funded by the National Engineering Research Center of UHV Technology and Novel Electrical Equipment Basis (NERCUHV-2023-KF-09).

Data Availability Statement: The original contributions presented in this study are included in the article. Further inquiries can be directed to the corresponding author.

Conflicts of Interest: Authors Wenbo Zhu, Mingli Fu, and Linjie Zhao were employed by the company Electric Power Research Institute of China Southern Power Grid. The remaining authors declare that the research was conducted in the absence of any commercial or financial relationships that could be construed as a potential conflict of interest.

References

1. Sonerud, B.; Mäkelä, T.; Josefsson, S. Design Evaluation of HVDC Flexible Joints using Numerical Simulations and Conductivity Measurements. In *Annual Report—Conference on Electrical Insulation and Dielectric Phenomena, CEIDP*; Institute of Electrical and Electronics Engineers Inc.: Piscataway, NJ, USA, 2020; pp. 447–450. [CrossRef]
2. Fernandez, J.; Guffond, R. Breakdown Prediction Modelling of XLPE Insulation under HVDC Thermoelectrical Stress. In *Annual Report—Conference on Electrical Insulation and Dielectric Phenomena, CEIDP*; Institute of Electrical and Electronics Engineers Inc.: Piscataway, NJ, USA, 2022; pp. 459–462. [CrossRef]
3. Fard, M.A.; Farrag, M.E.; McMeekin, S.; Reid, A. Electrical treeing in cable insulation under different HVDC operational conditions. *Energies* **2018**, *11*, 2406. [CrossRef]
4. Fournier, D.; Lamarre, L. Effect of Pressure and Temperature on Interfacial Breakdown between Two Dielectric Surfaces. In Proceedings of the 1992 Annual Report Conference on Electrical Insulation and Dielectric Phenomena (CEIDP), Victoria, BC, Canada, 18–21 October 1992; pp. 229–235. [CrossRef]
5. Tian, X.; Li, Y. Research on Electrical Characteristics of Insulating Silicone Oil Based on Polarization-Depolarization Current Method. In Proceedings of the 2020 IEEE 1st China International Youth Conference on Electrical Engineering (CIYCEE), Wuhan, China, 1–4 November 2020; IEEE: Piscataway, NJ, USA, 2020; pp. 1–6. [CrossRef]
6. Chen, Z.; Tang, J.; Luo, C.; Zhang, J. A Study on Aging Characteristics of Silicone Oil in HV Oil-filled Cable Termination Based on Infrared Thermal Imaging Test. *MATEC Web Conf.* **2015**, *22*, 05026. [CrossRef]

7. Li, J.; Han, C.; Du, B.; Takada, T. Deep trap sites suppressing space charge injection in polycyclic aromatic compounds doped XLPE composite. *IET Nanodielectrics* **2020**, *3*, 10–13. [CrossRef]
8. Zhang, C.C.; Li, Y.F.; Hu, M.Y.; Ma, F.L.; Zhao, H.; Han, B.Z. *Conductivity Properties of XLPE Insulation Used for HVDC Cable After Accelerated Thermal Ageing*; IEEE: Piscataway, NJ, USA, 2018.
9. Isabella, N.; Tim, B.; Marvin, B.; Bastian, S. Effect of Ageing on the Dielectric properties of EPDM for HVDC Cable Joints. In Proceedings of the VDE High Voltage Technology 2020; ETG-Symposium. [VDE Verlag], Online, 9–11 November 2020.
10. Xing, Y.; Guo, X.; Kong, F.; Pathiraja, K.; Wang, Y.; Li, J. Interfacial breakdown characteristics of XLPE/EPDM double-layer for DC cable terminations with different treatment conditions. *J. Phys. D Appl. Phys.* **2025**, *58*, 115503. [CrossRef]
11. Evangelou, C.; Zaky, A.A.; Megahed, I.Y. The effect of organic additives on the breakdown strength of transformer oil. *J. Phys. D Appl. Phys.* **1973**, *6*, 103. [CrossRef]
12. Angerer, L. Effect of organic additives on electrical breakdown in transformer oil and liquid paraffin. *Proc. Inst. Electr. Eng.* **1965**, *112*, 1025. [CrossRef]
13. Brian, S. *Fourier Transform Infrared Spectroscopy Fundamentals of Second Edition*; CRC Press, Taylor & Francis Group: Boca Raton, FL, USA, 2011.
14. Gong, Y.; Chen, X.; Wu, W. Application of fourier transform infrared (FTIR) spectroscopy in sample preparation: Material characterization and mechanism investigation. *Adv. Sample Prep.* **2024**, *11*, 100122. [CrossRef]
15. Arkles, B.; Launer, P.J. Infrared Analysis of Organsilicon Compounds: Spectra-Structure Correlations. Available online: https://www.researchgate.net/publication/273439829_Silicon_Compounds_Silanes_Silicones (accessed on 14 April 2025).
16. Ayalew, M.E. DFT Studies on Molecular Structure, Thermodynamics Parameters, HOMO-LUMO and Spectral Analysis of Pharmaceutical Compound Quinoline (Benzo[b]Pyridine). *J. Biophys. Chem.* **2022**, *13*, 29–42. [CrossRef]
17. Takada, T.; Miyake, H.; Tanaka, Y.; Yoshida, M. Quantum Chemical Calculation Studies on Interface Charge Transfer between Electrode and Polyethylene under Electrical Stress. In Proceedings of the 2014 International Symposium on Electrical Insulating Materials, Niigata, Japan, 1–5 June 2014.

energies

MDPI

Article

Influence of Green Algae on the Surface Wetting Characteristics of Porcelain Insulators

Xiaolai Li [1], Xiangdong Wu [1], Shiqiang Yang [1], Beichen Gao [1], Liang Li [2] and Bin Cao [2,*]

1 Ultra-High Voltage Company, State Grid Hubei Electric Power Co., Ltd., Wuhan 430050, China; lixl@hb.sgcc.com.cn (X.L.)
2 Tsinghua Shenzhen International Graduate School, Tsinghua University, Shenzhen 518055, China; liliang23@mails.tsinghua.edu.cn
* Correspondence: cao.bin@sz.tsinghua.edu.cn

Abstract: Insulator pollution flashover is a serious fault in power transmission systems, with surface wetting being a key prerequisite for its occurrence. The unique electrostatic properties of HVDC transmission networks promote pollution accumulation and alter wetting behavior. In southwest China's warm, humid mountainous regions, algae adhesion on DC insulators significantly affects surface wetting, increasing the risk of flashover under extreme weather conditions. This study proposes a surface-conductivity–based method to measure the water absorption of pollution layers on insulators. It quantitatively assesses the impact of algae on wetting characteristics, including saturated water absorption and salt loss rate. Experimental results show that in fog, saturated water absorption decreases as the tilt angle increases, while soluble salt content decreases over wetting time. NSDD has a minimal effect on saturated absorption. Moreover, the presence of algae significantly alters wetting behavior, increasing saturated water absorption by 27–47% and reducing salt loss. These findings provide an insight into the role of biological contamination in pollution flashover processes in high-humidity regions.

Keywords: porcelain insulator; wetting characteristics; green algae contamination; flashover; water absorption capacity; surface conductivity

1. Introduction

The distribution of energy resources in China shows a significant regional imbalance, with the western region having abundant energy but limited demand, while the eastern region has huge energy demand but relatively scarce resources. The use of a high-voltage direct current (HVDC) system, with its high efficiency and low loss transmission characteristics, has successfully transported the abundant energy in the west to the areas with strong demand in the east at a large scale, effectively alleviating the mismatch between energy distribution and demand. Now, China has established a "system of UHV transmission projects straights", which not only greatly facilitates the widespread consumption of clean energy, but also strongly guarantees China's energy security for crosses 19 and 20.

However, with the expansion of power system construction, the environment in which transmission lines are located is becoming more complex and diversified, and transmission lines are facing the notable contamination challenge of pollution (e.g., industrial dust, bird droppings, algae) [1]. In addition, currently, extreme climate events are intensifying globally, and ultra-rapid wetting of the pollution layer, caused by complex meteorological conditions such as large temperature differences, heavy rainfall, and short-term dense fog, frequently occurs. In mountainous areas with high humidity and micrometeorological environments, transmission lines are facing new challenges. Algae attached to the surface of a large number of insulators are becoming a new source of external insulation contamination. Field tests conducted using ultraviolet corona imaging on three towers of the ±800 kV Qishaoo HVDC transmission line demonstrated that compared to insulator strings with mild algae contamination exhibiting a discharge frequency of 1.3 times/min, heavily and severely contaminated insulator strings showed significantly increased discharge frequencies of 2.8 times/min and 4.5 times/min, respectively. Since the current insulation configurations in mountainous areas are generally designed according to the light contamination zone, these algae attachment phenomena pose a serious threat to the safe operation and maintenance of transmission lines. On the one hand, algae attachment accelerates pollution accumulation and increases its degree of attachment by enhancing hydrophilicity and surface roughness, coupled with water absorption and water retention of algae, resulting in partial discharges of insulators under low-ambient humidity conditions [2–5]. On the other hand, the current lack of countermeasures for algae management on the surface of transmission lines also leads to an increase in the number of outages for cleaning, while algae on the surface of insulators break easily under external forces, leading to an increased risk of accidents caused by hand-slip or foot-slip accidents by O&M personnel, which greatly affects the safety and efficiency of maintenance work [6–10].

The process of pollution flashover is generally divided into four stages as follows: surface accumulation of pollution, surface wetting, local discharge, and local arc development until flashover [11,12]. Among these, wetting is a necessary condition for insulator flashover, and algae attached to the insulator surface greatly affect the insulator wetting characteristics, thus increasing the probability of pollution flashover accidents. In addition, according to the Obennaus pollution flashover model, whether the surface arc can continue to develop and eventually penetrate the leakage distance of the insulator (i.e., cause a complete flashover) depends on the energy balance during the arc development process; that is, whether the sum of the arc partial voltage drop and the residual pollution layer voltage drop is less than the total voltage, and the conductivity of the pollution layer is the key factor affecting this process [11–13]. The surface conductivity of the insulator surface is, in turn, affected by the dissolution of soluble salts in the contamination layer [14,15]. Before the contaminated layer on the insulator surface reaches saturation wetting, the higher the degree of wetting of the contaminated layer, the lower the pollution flashover voltage of the insulator with the same level of contamination. However, when the insulator surface is saturated with wetting, the pollution flashover voltage of the insulator string does not increase significantly, and even decreases, due to the loss of the dissolved contaminated layer. Therefore, it is of great significance to study the contamination wetting characteristics of algae-covered insulators and the saturated water absorption of the contamination layer on the insulator surface after wetting, in order to reveal the influence of algae on the contamination layer wetting characteristics and the flashover characteristics of the insulator surface [16–22].

To this end, this paper first proposes a method for measuring insulator surface water absorption based on surface conductivity, and then takes porcelain insulators for station lines in an Enshi 500 kV substation as the test material to study the influence of algae attachment on the saturated water absorption of insulators, as well as the influence on the loss of pollution soluble salts. The test results provide a reference for studying the flashover process of insulators in foggy environments, bridging the gap between biological contamination and electrical insulation failure, thus addressing an emerging issue driven by climate change and ecological shifts in grid environments [23,24].

2. Test Methods

This study aims to investigate the influence of green algae growth and adhesion on the surface wetting characteristics and soluble salt loss properties of porcelain insulators by measuring the following two key indicators: saturated water absorption and the residual rate of soluble salts in the contamination layer. The surface conditions of the porcelain insulators were categorized as clean or algae-covered, leading to four experimental series, measuring the following: (1) the effect of green algae adhesion on saturated water absorption under varying parameters, and (2) the influence of green algae adhesion on the loss characteristics of soluble salts.

The test samples were selected from the porcelain insulators used in the Enshi 500 kV substation. To eliminate interference from the inherent contamination components (both soluble and insoluble) in the algae contamination layer and algal cells, all samples underwent a decontamination pretreatment with deionized water. The specific procedure was as follows: first, the insulator surfaces were rinsed with deionized water at a rate of 5 mL/cm^2, and the rinsing solution was collected. Then, the algae-covered insulators were immersed in deionized water for 12 h and subsequently air-dried. Finally, the rinsing process was repeated, and the rinsing solution was collected again. The major elemental contents in the rinsing solution before and after treatment were analyzed via ICP spectroscopy (Table 1). The results confirmed that the concentrations of major soluble salt cation elements (e.g., Na^+, K^+, Ca^{2+}, Mg^{2+}) were below 0.005 mg/L, indicating negligible contamination interference from the algae layer itself. The treated algae-covered samples were labeled as "algae-coated samples," while the thoroughly cleaned insulators of the same model were labeled as "blank samples." On both algae-coated and blank samples, approximately 120° fan-shaped test areas were demarcated using insulating tape, with copper-conductive tape applied radially at 10 cm intervals as electrodes (Figure 1).

Table 1. Content of main elements before and after treatment.

Types of Elements	Element Concentration (mg/L)	
	Before Rinsing	After Rinsing
K^+	2.495	<0.005
Ca^{2+}	0.928	<0.005
Na^+	19.560	<0.005
Mg^{2+}	0.177	<0.005

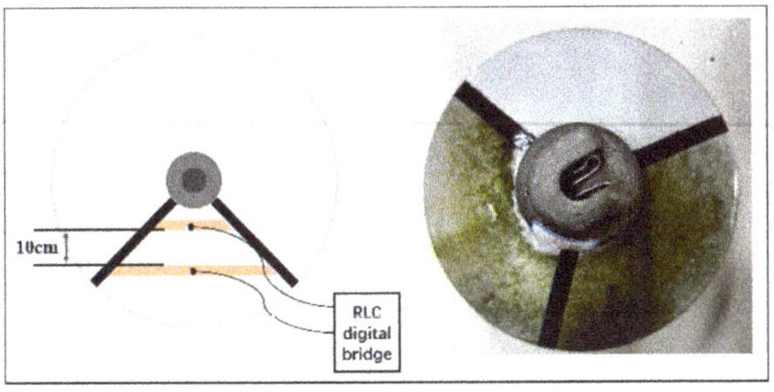

Figure 1. Schematic diagram of insulator processing sample.

Because the saturated moisture after the contamination of the layer of water absorption by the insulator is very low, it is difficult to use the weighing method to directly measure the surface contaminated layer of water absorption. For this reason, this study proposes a method to measure the water absorption of the insulator surface based on surface conductivity.

First, a NaCl solution with a specific conductivity is configured, and then, after the NaCl solution is converted into salt mist with the same conductivity using an ultrasonic atomizer, it is introduced into an artificial fog chamber to provide a wetting condition for the contamination layer. Then, an insulating specimen coated only with an insoluble substance (kaolin) is placed into the artificial fog chamber to be wetted. Tests conducted via the two-electrode method using an RLC bridge demonstrated that an increase in moisture exposure time caused the salt spray particles to adhere, and the conductivity of the contamination layer surface increased with the formation of a water film. A high-precision digital bridge was utilized to measure the change in surface conductivity during the wetting process. Since the volume conductivity of the contaminated layer solution is the same as that of the salt spray solution, the amount of water attached to the contaminated layer on the insulating surface can be calculated according to the relationship between the surface conductivity of the contaminated layer and its volume conductivity. A schematic diagram of the insulator contamination layer wetting process is shown in Figure 2.

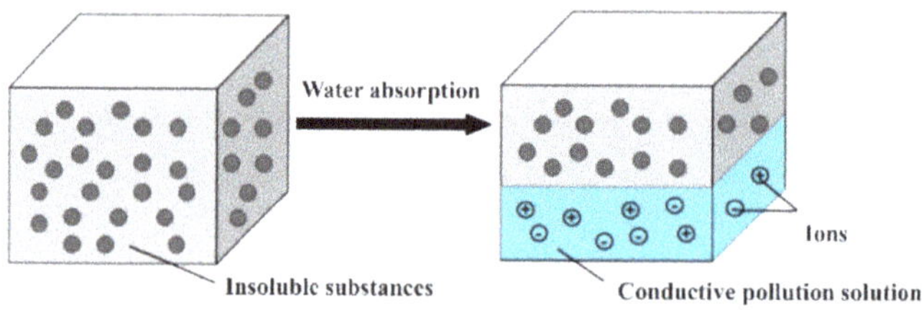

Figure 2. The wetting process of the contamination layers on the insulators.

During the humidification process, the surface resistance of the contamination layer is measured and recorded using an RLC bridge.

$$K = f\frac{1}{R} \tag{1}$$

where R is the measured surface resistance and f is the shape factor of the insulator surface. As shown in Equation (2), the measured conductivity can be corrected for temperature.

$$K_{20} = k_t \cdot K_t \tag{2}$$

where K_{20} is the corrected surface conductivity below 20 °C; K_t is the surface conductivity measured at a temperature of t °C; and k_t denotes the temperature coefficient [18]. The relationship between the surface conductivity and the salt spray conductivity is

$$K = \sigma \times h \tag{3}$$

where K denotes the surface conductivity; h denotes the thickness of the water film after wetting; and σ denotes the conductivity of the salt spray. On the other hand, water absorption m_{water} is calculated as

$$m_{water} = \rho_{water} \times S \times h \tag{4}$$

where ρ_{water} is water density and S is the surface area (1 cm^2). Figure 3 shows the relationship between surface conductivity and water absorption.

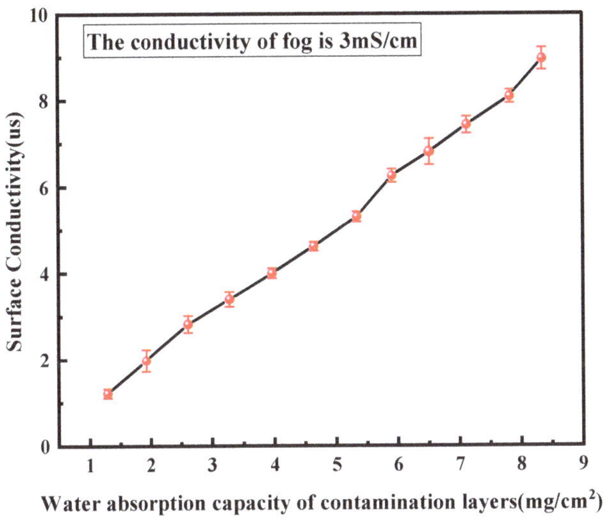

Figure 3. The relationship between surface conductivity and water absorption of contamination layer.

The test platform is mainly composed of a small fog chamber, an ultrasonic nebulizer, and an RLC bridge, in which an air heating line, an exhaust fan, and a temperature and humidity meter are mainly used for control and detection of environmental parameters, such as temperature and fog concentration, inside the artificial fog chamber, as shown in Figure 4. A test tilt-adjustable bracket was placed inside the fog chamber. Before the test, we controlled the humidity of the artificial fog chamber to 93%RH, and the average fog water settling rate in the fog chamber was measured by gravimetric sampling using a fog cup, which measured 0.245 mg/cm^2·min (±0.02 mg/cm^2), and the uniformity was verified by measuring the conductivity at four locations in the artificial fog chamber (deviation < 5%). In addition, the spray distance (40 cm) was optimized to ensure that droplet sizes (10–20 μm) adhered to IEC 60,507 for artificial pollution tests.

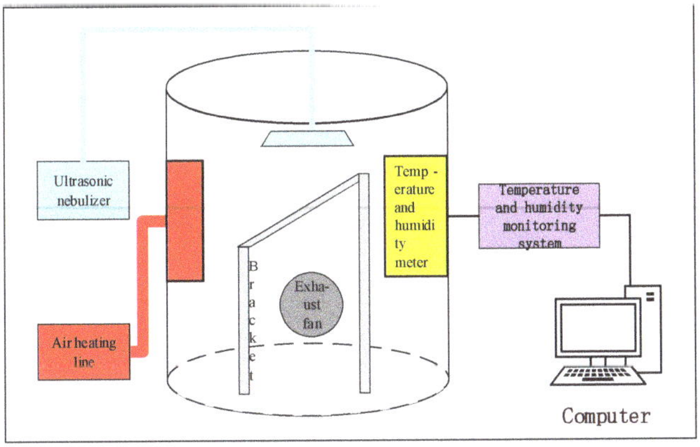

Figure 4. Test platform.

The specific experimental procedure was as follows: first, an artificial contamination layer containing only kaolin was applied to the sample surfaces using the solid-layer method. The samples were then placed in a drying oven and dried at 40 °C for 3 h to completely dry the contamination layer [25,26]. For testing, the test specimen was moved into an artificial fog chamber, and the spray position and angle were adjusted so that the vertical distance from the surface of the test specimen to the sprayer was 40 cm. A NaCl solution with a specific conductivity was prepared at 20 °C and then added to the ultrasonic nebulizer, and the salt mist was sprayed into the fog chamber. During the wetting process, the surface conductivity trend was measured and recorded by an RLC bridge.

In the experiment investigating the influence of green algae adhesion on soluble salt loss characteristics, artificial contamination with NaCl and kaolin was first applied to both algae-coated samples and blank samples using the solid-layer method. The contamination was proportionally prepared according to the test area dimensions to achieve an equivalent salt deposit density (ESDD) of 0.5 mg/cm^2 and a non-soluble deposit density (NSDD) of 1 mg/cm^2 in the test areas. In addition, due to insulator contamination in the wet conditions of the loss of a slow process, in order to avoid insulators being in the artificial fog chamber in the wet time for too long, before the test, an equal amount of deionized water was sprayed on to the surface of the blank samples and the samples that were algae-coated with a spray can in order to accelerate the wetting. After the insulator was wetted, it was placed horizontally into the artificial fog chamber, and then the clean fog generated only by deionized water was passed into the artificial fog chamber. Due to the existence of the insulator umbrella, itself at an angle of inclination, the soluble salts were constantly dissolved and lost through the edge. During the test, a cotton swab was dipped into the edge of the insulator test area and dissolved into 10 mL of deionized water to determine the conductivity, which was recorded every 60 s.

Conductivity–time curves were fitted to the conductivity recorded at each discrete time.

$$\sigma = f(t) \tag{5}$$

The integration of the curve yields a total loss soluble salt conductivity–time curve as follows:

$$F(t) = \int_0^t \sigma \mathrm{d}t = \int_0^t f(u)\mathrm{d}t \tag{6}$$

where $F(t)$ is the conductivity corresponding to the total amount of soluble salts lost from the insulator surface dissolved into 10 mL of water up to the moment t, and $F(\infty)$ is the conductivity corresponding to the dissolution of all the soluble salts on the insulator surface dissolved into 10 mL of water. Since the amount of pollution solution sampled each time is very small, dissolving it into 10 mL of deionized water does not lead to the phenomenon in which the dissolution of soluble salts reaches saturation, and the conductivity of the test solution is proportional to the number of soluble salts in the pollution solution.

In order to eliminate the interference of intracellular salt release after algae cell rupture, the soluble salt residual rate $G(t)$ of artificial pollution and algae pollution was compared and analyzed.

For algae-coated samples,

$$G_1(t) = \frac{F_1(\infty) - F_1(t)}{F_1(\infty)} \times 100\% \tag{7}$$

For blank samples,

$$G_2(t) = \frac{F_2(\infty) - F_2(t)}{F_2(\infty)} \times 100\% \tag{8}$$

In order to ensure the accuracy of the test, in the above tests, five test samples were selected to repeat the test, fit the data after the test, remove the curve with the largest dispersion, and then take the average of the valid data at each sampling point to fit the final curve.

3. Test Results and Analysis

3.1. Trends in the Conductivity of the Contamination Layer

The surface conductivity of the blank samples during the wetting process was measured using a high-precision bridge. The surface conductivity trend of the test samples placed at a tilt angle of 45° during the wetting process is shown in Figure 5.

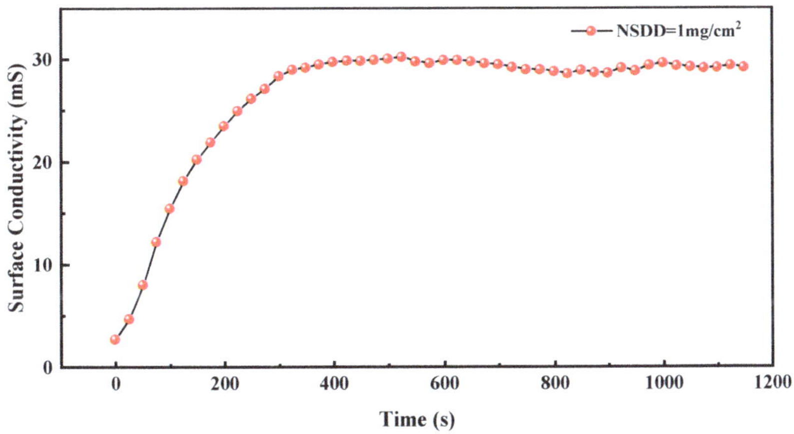

Figure 5. Surface conductivity of the test specimen with 45° inclination degree during the wetting time.

As shown in Figure 5, the surface conductivity of the test sample increases gradually as the wetting time increases, and finally reaches the maximum value gradually. The change in surface conductivity can be divided into two stages.

In the first stage, as the wetting time increases, salt spray particles are gradually deposited on the contamination layer, the amount of water attached to the contamination layer and the number of conductive particles in the contamination layer are synchronized to increase in a certain ratio, and the surface conductivity of the samples gradually increases. Since the accumulation rate of salt spray on the surface of the sample is greater than the rate of salt spray loss, the water absorption of the contamination layers increases. In the second stage, as the degree of wetting increases, the adhesion rate of salt spray particles and the loss of salt spray under gravity reach equilibrium. At this time, the sample surface conductivity reaches a stable value because the water absorption of the contamination layer no longer increases.

3.2. Influence of NSDD on the Amount of Water Absorbed by the Contamination Layer

According to the above relationship between surface conductivity and water absorption in the contamination layer, the saturated water absorption can be converted by the maximum surface conductivity during wetting, and the relationship between saturated water absorption and NSDD is shown in Figure 6.

When the insulator surface attaches to a layer of insoluble material in the insulating surface to form many loose porous structures, these loose structures can be adsorbed and stored in a certain amount of water, making the saturated layer of contamination increase with the amount of water and with the increase in the contaminated layer gray density.

As shown in Figure 6, for the vertically arranged test samples, the saturated water absorption of the contamination layer was in the range of 1.68–2.7 mg/cm^2 when the NSDD was in the range of 0.5–4.0 mg/cm^2. However, for the specimen with an inclination of 45°, the saturated water absorption of the soiled layer was in the range of 3.04–3.42 mg/cm^2. The data show that the saturated water absorption of the 45°-inclined blank sample (3.04–3.42 mg/cm^2) is significantly higher than that of the vertical sample (1.68–2.7 mg/cm^2). At the same inclination angle, an increase in NSDD leads to an increase in the water absorption of the contaminated layer on the insulator surface.

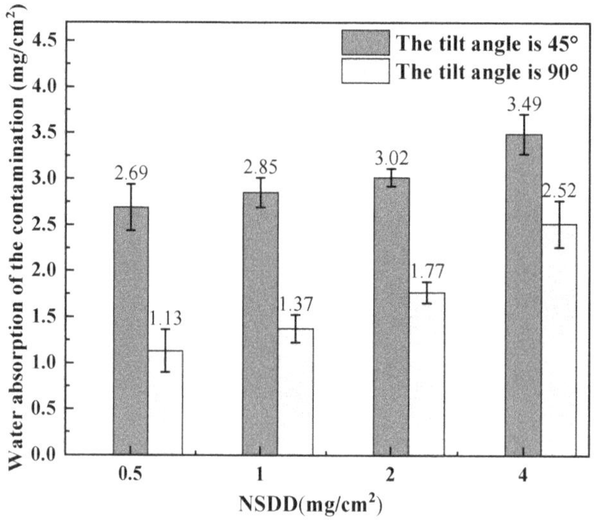

Figure 6. Saturated water absorption of the contamination layers with different NSDDs.

3.3. *Influence of Tilt Angle on the Amount of Water Absorbed by the Soiled Layer*

Changing the angle of the blank samples, the test was carried out at different inclination angles and at the same NSDD; the test results are shown in Figure 7. Through this, it can be seen that the water absorption on the contamination layer decreases as the tilt angle increases. When the test piece is ideally placed horizontally, it is obvious that the salt spray gradually accumulates on the insulator surface in the fog chamber, the thickness of the water film keeps increasing with time, and the saturation water absorption is infinite.

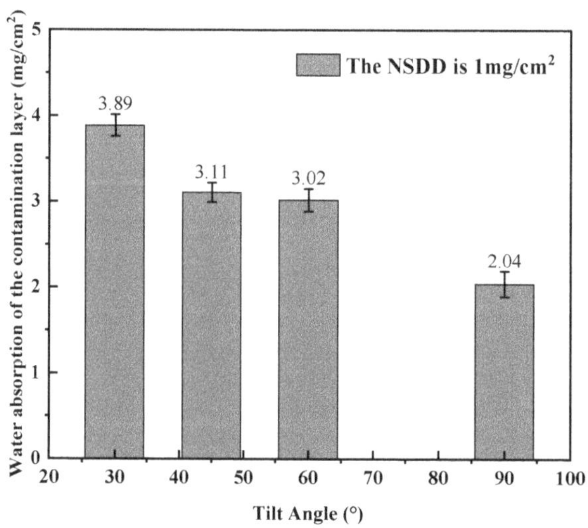

Figure 7. Saturated water absorption of the contamination layers with different inclination angles.

4. Effect of Algae on the Wetting Characteristics of Insulators

4.1. Loss of Soluble Salts from the Contamination Layer

The loss characteristics of the soluble salts were compared with the insulator surface with and without algae overgrowth on the surface. As shown in Figure 8, the soluble salt's residual amount of algae-coated sample is larger than that of the blank sample during the whole process of wetting; from the loss rate, the soluble salt residual rate of insulator surface contamination ranges from 100% to 60%, and the time required for blank sample and algae-coated sample is not significantly different, while the soluble salt residual rate is reduced to 20%, and the area of the blank sample reaches about 500 s, while the algae-coated sample reaches about 2000 s; from the final contamination soluble salt residual rate, under the condition of a sufficiently long wetting time, the soluble salt residual rate of the blank sample can be reduced to 20%, and the algae-coated sample can reach about 2000 s. At 2000 s, from the final contamination soluble salt residual rate, when the moisture time is long enough, the blank sample soluble salt residual rate is finally around 0.5%, while the algae-coated sample soluble salt residual rate is finally around 2%.

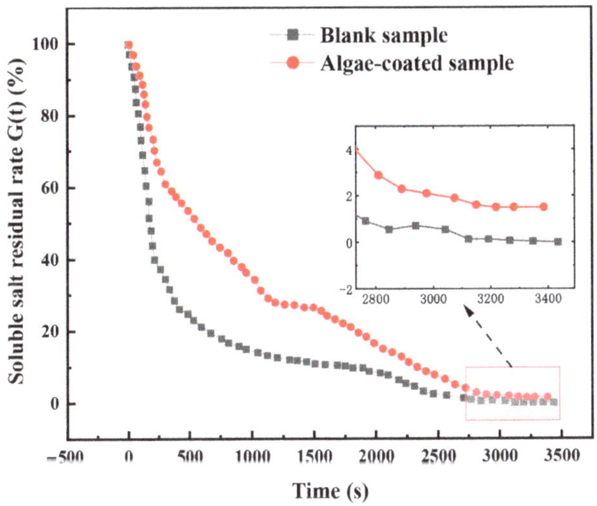

Figure 8. Soluble salt residual rate curve of the contamination layer.

According to the change in loss rate, the soluble salt loss process can be divided into a rapid loss period (0–200 s), a slow loss period (200–1000 s), and a stable loss period (>1000 s). During the rapid loss period, the blank sample and algae-coated sample in the more easily soluble salt show a faster rate of loss; during the slow loss period, due to the attachment of algae, the algae-coated sample region show a change in the roughness of the insulator surface, resulting in the algae filtration region of the soluble salt showing a slower rate of loss. Meanwhile, at this stage, the blank sample region of the soluble salt remains at a faster rate of loss; during the stable loss period, the artificial filtration region and algae filtration region of the soluble salt are nearly all lost, but the algae filtration soluble salt notably shows a loss in the third stage of the stable loss period. During the stable loss period, the soluble salts in both the artificial and algae-coated sample areas were almost completely lost, but the soluble salt residual rate of the algae-coated sample was stabilized at a relatively high value, which may be due to the fact that the intracellular constituents of the algae cells could only be solubilized after the rupture of the cells during the final scrubbing.

4.2. Influence of Algae on the Saturated Water Absorption of the Contamination Layer

To study the influence of algae on the saturated water absorption of the insulator's contaminated layer, the water absorption measurement method for the contaminated layer based on surface conductivity described in Section 1 was used.

From Figure 9, it can be seen that the saturated water absorption of the algae-coated samples layer is larger than that of the blank samples layer at different inclination angles. The saturated water absorption of both types of contamination in the salt spray environment decreases as the tilt angle increases, but there are differences in their decreasing trend. With an increase in tilt angle from 30° to 90°, the saturated water absorption of blank samples decreased by 1.80 mg/cm^2, down by 47.37%; while the saturated water absorption of algae-coated samples decreased by 1.595 mg/cm^2, down by 27.88%. The saturated water absorption of the blank samples was more highly affected by the tilt angle.

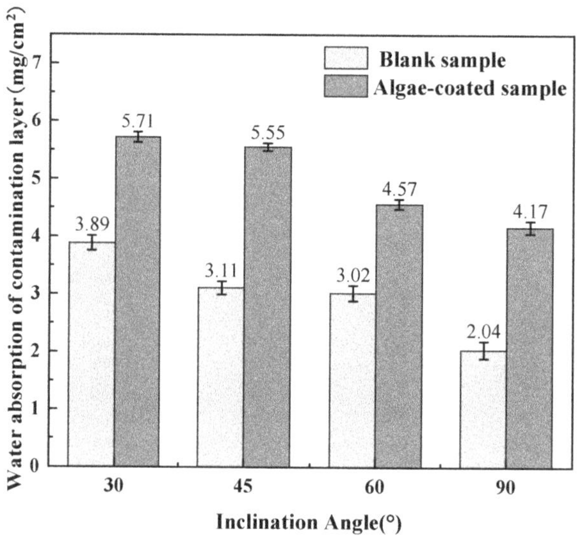

Figure 9. Comparison of water absorption of the contamination layer at different inclination angles.

The reason for the above phenomenon is that algae-coated samples usually grow by forming a tighter structure through aggregation between cells, and this structure forms a large number of small pores and voids that can trap and hold water. In contrast, the insoluble matter in general contamination forms many loose porous structures mainly on the insulating surface, and these loose structures can trap and store a certain amount of water [27]. Compared with the large number of voids that exist between algae cells, the water absorption capacity of the loose small pores in the ash is weaker, so the water absorption capacity of the algae-coated samples is stronger. In addition, after the algae attaches to the surface of porcelain insulators, it changes the original smooth surface structure and greatly enhances the roughness of the surface, and under a certain tilt angle, the rough surface does not easily facilitate the flow and loss of the water film; therefore, the saturated water absorption of the algae-coated sample layer is affected by the tilt angle to a small extent.

5. Conclusions

This study presents a method for analyzing the water absorption of contamination layers on insulator surfaces based on surface conductivity. Using this method, the wetting characteristics of artificial and algae-induced pollution were compared, and the algae's roles in inhibiting salt dissolution and enhancing water retention were quantitatively assessed.

The test results show that for samples without algae, both NSDD and tilt angle influence saturated water absorption. For vertically aligned specimens, the saturated water absorption ranges from 1.68 to 2.7 mg/cm^2, and is strongly affected by NSDD. However, when tilted (e.g., at 45°), the absorption becomes largely independent of NSDD, ranging narrowly between 3.04 and 3.42 mg/cm^2.

Algae contamination was found to reduce the soluble salt loss rate by 40% by increasing surface roughness. Artificially polluted surfaces reached a 20% residual salt rate in 500 s, whereas algae-contaminated surfaces required 2000 s. Across all tilt angles, algae-coated samples exhibited consistently higher saturated water absorption than clean samples. As the tilt angle increased from 30° to 90°, the saturated water absorption in blank samples decreased by 47.37%, compared to 27.88% in algae-coated samples, indicating greater sensitivity to tilt in non-algae-contaminated surfaces.

This study highlights the significant impact of algae on the wetting behavior of porcelain insulators under rapid-wetting conditions. However, our results are limited to UHVDC-grade porcelain insulators in artificial fog environments. Future work will examine behavior under medium- and low-humidity conditions and extend the analysis to composite insulator materials with hydrophobicity.

Author Contributions: Conceptualization, X.L., X.W., L.L. and B.C.; Methodology, X.L., X.W., S.Y. and B.G.; Validation, S.Y. and B.C.; Investigation, B.G., L.L. and B.C.; Resources, X.L.; Data curation, X.L. and S.Y.; Writing—original draft, L.L.; Writing—review & editing, B.G. and B.C. All authors have read and agreed to the published version of the manuscript.

Funding: This work was supported by the Ultra-High Voltage Company, State Grid Hubei Electric Power Co., Ltd. Science and Technology Project "Treatment of algae on insulator surfaces in 2024" (30 June 2024).

Data Availability Statement: The data that support the findings of this study are available from the corresponding author upon reasonable request.

Conflicts of Interest: Authors Xiaolai Li, Xiangdong Wu, Shiqiang Yang and Beichen Gao were employed by the Ultra-High Voltage Company, State Grid Hubei Electric Power Co., Ltd. The remaining authors declare that the research was conducted in the absence of any commercial or financial relationships that could be construed as a potential conflict of interest. The authors declare that this study received funding from Ultra-High Voltage Company, State Grid Hubei Electric Power Co., Ltd. The funder was not involved in the study design, collection, analysis, interpretation of data, the writing of this article or the decision to submit it for publication.

References

1. Zhang, D.D.; Liu, X.N.; Huang, X.N.; Zhan, Z.J. Pollution constitutes and flashover characteristics of composite insulators in typical industrial dust areas. *Electr. Power Eng. Technol.* **2022**, *41*, 162–168.
2. Ouyang, X.G. Study on the Growth Characteristics of Algae on Silicone Rubber Surface and its Influence on Electrical Properties. Ph.D. Thesis, Tsinghua University, Beijing, China, 2019.
3. Yang, S.F. Study on the Hydrophobicity of Algae Contaminated Silicone Rubber and its Prevention and Control Measures. Ph.D. Thesis, Tsinghua University, Beijing, China, 2020.
4. Tan, J.Y. Research on the Growth Characteristics of Algae on Silicone Rubber Surface. Master's Thesis, Tsinghua University, Beijing, China, 2022.
5. Jia, Z.D.; Wang, Y.; Ouyang, X.G.; Zhang, X.H.; Bai, H.; Chen, H.B. Influence of Green Algae on Hydrophobic Property of HTV Silicon Rubber. *High Volt. Eng.* **2017**, *43*, 885–890. [CrossRef]
6. Jia, Z.D.; Zeng, Z.; Chen, C.; Yang, C.X.; Wang, X.L.; Yan, Z.H. Properties of composite insulating materials under special contamination. *High Volt. Eng.* **2016**, *42*, 885–892. [CrossRef]

7. Zhu, L.W.; Cao, B.; Fu, X.Y.; Li, H.; He, L.; Mei, H.W. Study on the influence of uneven distribution of hydrophobicity along insulator string on pollution flashover characteristics. *High Volt. Appar.* **2019**, *55*, 193–197. [CrossRef]
8. Zhang, Z.J.; Liu, H.Z.; Wang, X.G.; Li, X.G.; Li, J.; Fu, M. Research status of external insulation flashover of EHV and UHV transmission lines. *Shandong Electr. Power* **2015**, *42*, 1–6+11.
9. Hu, Y.C.; Gao, T.H.; An, Y.Z.; Liu, X.H.; Yin, B.; Xian, R.C. Study on the influence of algae epiphyte on the electric field distribution of 35kV insulator type voltage limiter. *Insul. Surge Arrester* **2020**, *6*, 102–108. [CrossRef]
10. Gao, T.H.; Hu, Y.C.; An, Y.Z.; Xian, R.C.; Hou, F. Study on the influence of algae on the electric field distribution of 110 kV composite insulators. *Insul. Mater.* **2019**, *52*, 59–64. [CrossRef]
11. Liang, X.D.; Chang, C.Y.; Zhou, Y.X. *High Voltage Engineering*; Tsinghua University Press: Beijing, China, 2003; pp. 75–76.
12. Su, Z.Y. The fundamental way out to prevent large area pollution is to improve the basic external insulation level of the power grid reflection on the large-scale dirty accident in my country's power grid. China Electricity. *China Power* **2003**, *12*, 60–64.
13. Rizk, A.M. A criterion for AC Flashover of Polluted Insulators. In Proceedings of the IEEE Power Engineering Soc. Winter Power Meeting, New York, NY, USA, 31 January–5 February 1971.
14. Ye, L.A.; Zhou, W.J.; Yang, S.; Li, L.N.; Yu, J.H. Influence of dirt on insulator string on voltage distribution. *High Volt. Electr. Appl.* **2015**, *51*, 103–108. [CrossRef]
15. Mei, H.W. Wetting Characteristics and Surface Discharge Suppression Measures of Contaminated Insulator. Ph.D. Thesis, Tsinghua University, Beijing, China, 2011.
16. Li, C.H. Research on the Influence of Green Algae on the External Insulation of Composite Insulators. Master's Thesis, South China University of Technology, Guangzhou, China, 2018.
17. Liang, X.D.; Wu, C.; Yao, X.M.; Liu, Y.Y.; Liu, Y.F.; Wang, J.F. Effective Equivalent Salt Deposit Density of Polluted Silicone Rubber Insulators in Wetting Process. In Proceedings of the 2014 IEEE Conference on Electrical Insulation and Dielectric Phenomena (CEIDP), Des Moines, IA, USA, 19–22 October 2014; pp. 271–274. [CrossRef]
18. Wang, L.; Cao, B.; Mei, H.; Zhao, C.; Guan, Z. Effects of natural contamination components on the surface conductivity under saturated moisture. *IEEE Trans. Dielectr. Electr. Insul.* **2017**, *24*, 2945–2951. [CrossRef]
19. Mei, H.W.; Zhao, C.L.; Dai, H.Q.; Wang, L.M.; Guan, Z.C.; Zhang, F.Z. Study on wetting characteristics of polluted porcelain and glass insulators. *Proc. CSEE* **2014**, *34*, 1471–1480. [CrossRef]
20. Zhang, X.; Dai, D.B. Research on factors affecting saturation and wetness of insulator contamination layer. *High Volt. Appar.* **2013**, *49*, 123–128. [CrossRef]
21. Yu, H.Y.; Kong, Z.Z. Investigation of External Insulation Selection for Power Delivery & Transmission Equipment in Foggy Area. *Smart Power* **2000**, *5*, 13–16+2.
22. Jiang, X.L.; Shu, L.C.; Sun, C.X. *Power System Pollution and Ice Insulation*; China Electric Power Press: Beijing, China, 2009.
23. Lin, F.Y. Discussion on the measurement method of equivalent salt of polluted insulators. *Insul. Surge Arresters* **1979**, *3*, 24–31. [CrossRef]
24. Huang, D.C.; Liu, X.B.; Xu, T.; Ruan, J.J.; Wan, X.D. Study on saturated water content of pollution layer on the surface of steam mist suspension insulator. *High Volt. Eng.* **2014**, *40*, 3349–3356. [CrossRef]
25. *IEC 60507-2013*; Artificial Pollution Tests on High Voltage Insulators to be Used on AC Systems. IEC: Geneva, Switzerland, 2013.
26. *IEEE Std. 4-2013*; IEEE Standard Techniques for High-Voltage Testing. IEEE: New York, NJ, USA, 2013.
27. Wei, X.; Zhou, Z.C.; Zhao, C.L.; Gao, S.; Mei, H.W.; Wang, L.M. Influence of soluble salt components on the evaluation of insulator surface contamination. *High Volt. Eng.* **2013**, *39*, 3109–3115.

energies

MDPI

Article

Pressure-Dependent Breakdown Voltage in SF_6/Epoxy Resin Insulation Systems: Electric Field Enhancement Mechanisms and Interfacial Synergy

Lin Liu [1,2,*], Qiaogen Zhang [3], Xiangyang Peng [1], Xiaoang Li [3], Zheng Wang [1] and Shihu Yu [1]

1 Electric Power Research Institute of Guangdong Power Grid Co., Ltd., Guangzhou 510080, China
2 School of Electric Power Engineering, South China University of Technology, Guangzhou 510641, China
3 School of Electrical Engineering, Xi'an Jiaotong University, Xi'an 710049, China
* Correspondence: hvliulin@163.com; Tel.: +86-13822111715

Abstract: In SF_6 gas-insulated equipment, solid dielectrics critically degrade insulation performance by reducing the electric field's ability to withstand gas gaps. To investigate the critical role played by solid dielectric surfaces during the initial phase of gas–solid interface discharge phenomena, this paper experimentally measures the AC breakdown voltage (U_{bd}) of both dielectric surface-initiated breakdown (DIBD) and electrode surface-initiated breakdown (EIBD) across eight types of post insulator samples. Tests are conducted in 36 mm SF_6 gas gaps under pressures ranging from 0.1 to 0.4 MPa. Combined with electrostatic field simulations, the results reveal that DIBD requires substantially lower U_{bd} than EIBD under comparable maximum electric field (E_{max}) conditions. As gas pressure increases, this difference becomes more pronounced. This phenomenon can be explained by three key mechanisms: First, due to the regulatory effect of dielectric materials and shielding electrodes on the electric field distribution, the high-electric-field zone along the gas–solid interface exhibits a longer effective discharge path compared to that in a pure gas gap. This configuration creates more favorable conditions for discharge initiation and subsequent propagation toward the opposite electrode. Second, microscopic irregularities on the dielectric surface induce stronger local electric field enhancement than comparable features on metallic electrodes. Third, in high-electric-field regions adjacent to the dielectric surface, desorption processes significantly enhance electron multiplication during gas discharge, and this enhancement effect becomes more pronounced as gas pressure increases, further lowering the discharge inception threshold. As a result, discharge initiation at dielectric interfaces requires less stringent electric field conditions compared to breakdown in a gas gap, especially at high gas pressure. This conclusion not only accounts for the saturation behavior in the U_{bd}-p characteristic of SF_6 gas–solid interface discharges but also explains why surface contaminants/defects disproportionately degrade interfacial insulation performance relative to their impact on gas gaps.

Keywords: SF_6; gas–solid interface; insulator; surface flashover; gas discharge; electric field strength

1. Introduction

Due to its excellent insulating and arc-quenching properties, SF_6 gas-insulated metal-enclosed switchgears (GISs) and SF_6 gas-insulated transmission lines (GILs) are widely used in modern power systems [1–4]. Epoxy resin combines key advantages such as high insulation strength, high mechanical strength, ease of molding, high thermal endurance,

https://doi.org/10.3390/en18185014

and resistance to environmental aging. As a result, it has become an irreplaceable insulating and structural material in power equipment [4,5]. Specifically, in GIS/GIL systems, epoxy-based basin insulators and post insulators play crucial roles in providing both electrical insulation and mechanical support. However, the surface discharge along the gas–solid interface also introduces the risk of insulation failure. Surface flashover caused by over-voltage, metal particles, surface charge accumulation, surface cracks, and other foreign matters or defects is the main form of fault that threatens GIS/GIL insulation systems [6–10]. This fault type requires a long maintenance time, causes significant economic losses, and has a negative social impact. Therefore, investigating discharge mechanisms at gas–solid interfaces and optimizing insulator electric field configurations remain critical research priorities in high-voltage insulation engineering.

Previous studies have explored the effects of dielectric and metal surfaces on the initiation and development of gaseous discharge. In 1987, Verhaart et al. [11] highlighted the importance of photoemission effects for surface streamer criterion, which was drawn from their experimental finding that the critical electric field strength for surface streamer discharge at an SF_6-PTFE gas–solid interface is lower than that in a pure SF6 gas gap. In 2003, Jorgenson et al. [12] found that the photoemission process on solid surfaces promotes the development of surface streamers: the number of electrons generated by photoemissions on a solid surface is 6% greater than that captured in N_2 with a pressure lower than 0.1 MPa through simulation and comparative experiments. Chvyreva et al. [13,14] established that under AC conditions, the Raether–Meek criterion for streamer inception at electrodeless gas–solid interfaces exhibits gas-dependent behavior: gases with higher electron affinity gases require greater electron accumulation for avalanche-to-streamer transition. Our previous experimental observations revealed that in 0.06 MPa SF_6, the solid surface demonstrates pronounced adsorption effects on surface-creeping discharges, influenced by both discharge stochasticity and electrostatic forces. These studies indicate that the participation of solid dielectrics promotes initial surface discharge. However, none of the above studies could clarify the differences between the initial discharge on the solid surfaces versus electrode surfaces in GISs/GILs under high-pressure SF_6 gas discharge conditions.

In this paper, epoxy resin insulators with adjustable inner and outer electrodes for electric field regulation were employed to simulate the electrode surface-initiated breakdown (EIBD) and the dielectric surface-initiated breakdown (DIBD) under AC voltage in SF_6. The influence of the maximum system electric field strength and the gas pressure on the surface breakdown voltage were studied and discussed.

2. Test Platform and Samples

2.1. Test Platform

The test platform used in the experiments is depicted in Figure 1. The AC 50 Hz voltage supply was used to power the setup. The voltage was generated by a variable auto-transformer that is adjustable in the range 0–380 V. It was then directed to a high-voltage gas-insulated transformer (with an output voltage of up to 500 kV) and applied to the samples through water resistance. The function of water resistance is to limit the breakdown current and protect the surface of insulators from severe erosion. Referring to the contents of IEC 60296-2020 and IEC 60243-1:2013 standards [15,16], this paper adopts the fast boosting method, which involves raising the power frequency voltage to 50% of the estimated breakdown voltage within 10 s, and then using 2% of the estimated breakdown voltage as the boosting rate to make the insulator surface break down within 1 min. Since the duration of the applied voltage in the experiment did not exceed 1 min, the influence of thermal effects on the discharge process was not considered in this study [5]. The voltage waveform was measured by an oscilloscope (DPO4104, Tektronix China Co., Ltd., Shanghai,

China) with a bandwidth of 1 GHz and a sampling rate of 5 GHz, through measuring winding with a voltage ratio of 1000:1. The amplitude of the applied voltage is expressed here as a root-mean-square (RMS).

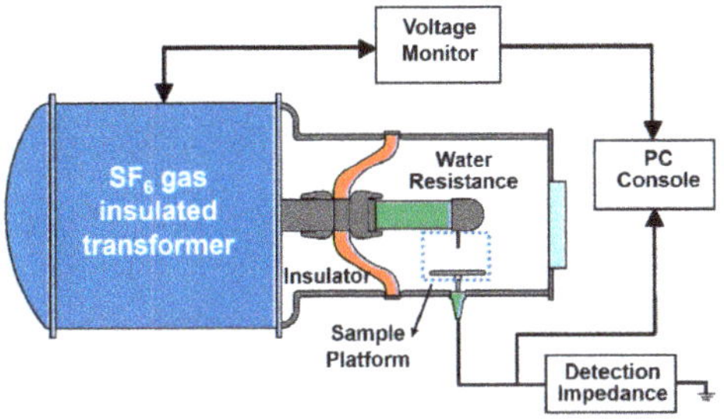

Figure 1. Power frequency surface flashover voltage testing platform for post insulators in SF_6 gas.

Following breakdown testing, all insulator samples underwent surface preparation which consisted of mechanical polishing with 2000-grit sandpaper (3M™) and ultrasonic cleaning in absolute ethanol. The surface roughness of the polished samples, as measured by the 3D optical profilometer (Up S-Dual 3D, RTEC Instruments, San Jose, CA, USA), was determined to be $R_a < 0.5$ μm and $R_z < 3$ μm, with occasional peak values (R_t) reaching up to 20 μm. To prevent the influence of discharge ablation traces on surface flashover voltage, the insulator sample was replaced after each breakdown. The used samples were polished with sandpaper before reinstatement for reuse. Insulator samples with severe surface ablation or carbonization that could affect subsequent measurements were discarded. The breakdown voltage (U_{bd}) was defined as the average of six valid measurements.

The test chamber was evacuated to a vacuum (gas pressure < 60 Pa), then filled with commercial SF_6 (purity > 99.997%) at 0.1 MPa. The gas pressure was monitored with a digital pressure sensor (Panasonic® DP-102, Osaka, Japan). All tests were conducted in a climate-controlled laboratory at 25 ± 2 °C and 40–60% RH.

2.2. Samples

The insulator samples were made of epoxy resin doped with Al_2O_3 fillers, and were manufactured using the vacuum casting process. Three different surface structures (concave, wave, and cylinder) as well as two shielding electrode structures (rod–plate outer shielding electrodes with small inner shielding electrodes and plate–plate outer shielding electrodes with big inner shielding electrodes) were employed to regulate the electric field distribution. The height of the epoxy resin body was 30 mm, and the height of the embedded electrode's exposed outer section was 3 mm; that is, the height of the gas gap was 36 mm, and the arc length along the insulator surface was 44 mm. Figure 2 presents the structure of the insulator samples.

Taking the gas gap composed of two types of external shielding electrodes into consideration, the numbering convention and electric field distribution of eight samples are shown in Figure 3. The electric field distribution of samples was calculated by the finite element method. Metal electrodes were modeled as equipotential conductor boundaries, and the relative permittivity of SF_6 gas and epoxy resin was set as 1.002 and 5.0, respectively. The applied voltage was set to 1000 V. A refined triangular free-mesh division achieved convergent computational results at 14,825 elements. Due to the simplicity of the geometric structure, the default solver settings in the software were sufficient to meet the computational requirements, and no specific additional configurations were necessary. The electric field distribution of the cross-section and gas–solid interface of the insulator sample is shown in Figure 3. Obviously, the E_{max} of samples (a) to (e) is located at the electrode surface while that of sample (f) to (h) is located at the insulator surface.

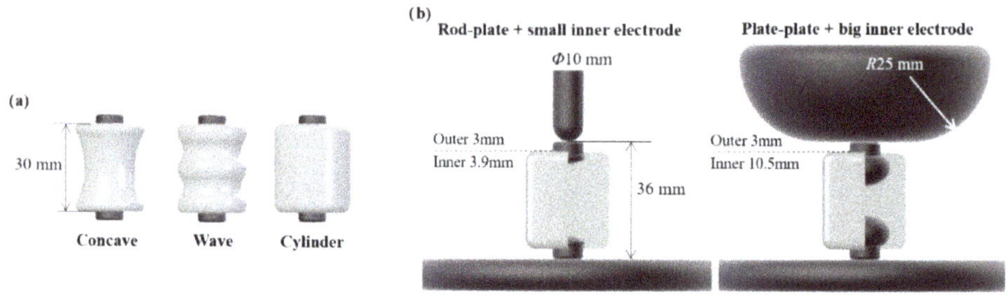

Figure 2. Structure of insulator samples. (**a**) The post insulator, showing concave, wave, and cylinder configurations. All are made of epoxy with a height of 30 mm. The diameters of the concave and wavy insulators are 24 mm, and that of the cylindrical one is 27 mm. (**b**) The shielding electrodes, comprising two types: a rod-plate outer electrode with a small inner shielding, and a plate-plate outer electrode with a large inner shielding. The rod radius is 5 mm. The embedding heights of the small and large inner shielding are 3.9 mm and 10.5 mm, respectively. Both inner shielding components have an exposure height of 3 mm.

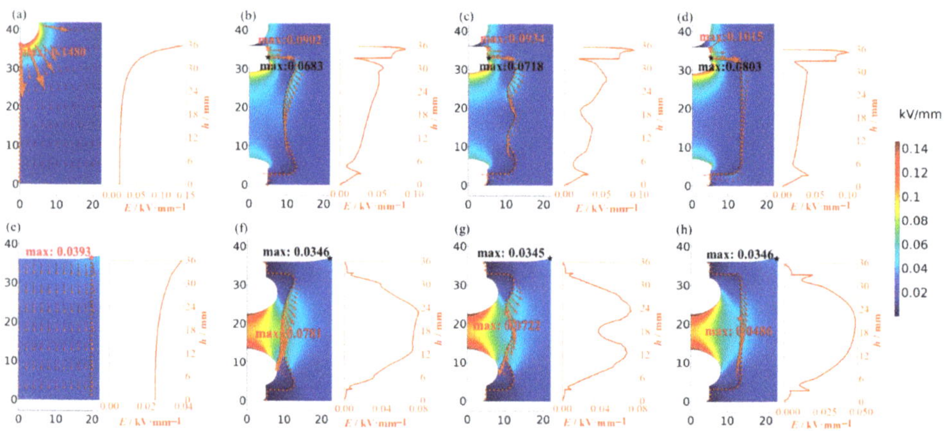

Figure 3. Electric field distribution in the insulator samples under an applied voltage of 1000 V, The size and direction of the red arrows in the electric field contour represent the strength and direction of the local electric field, respectively. (**a**) Gas gap with rod–plate shielding; (**b**) Concave insulator with rod–plate shielding; (**c**) Wave insulator with rod–plate shielding; (**d**) Cylinder insulator sample with rod–plate shielding; (**e**) Gas gap with plate–plate shielding; (**f**) Concave insulator sample with plate–plate shielding (**g**) Wave insulator sample with plate–plate shielding; (**h**) Cylinder insulator sample with plate–plate shielding.

Previous research has established that the breakdown process in compressed SF_6 follows a sequential progression: (1) initial electron generation triggering streamer inception and corona formation, (2) pre-breakdown partial discharge (PD) development, and (3) stepped leader propagation across the gap, ultimately culminating in arc breakdown [17–23]. As is well known, although SF_6 gas theoretically has excellent insulation strength, its insulation strength can drop sharply under extremely non-uniform electric field structures [22,23]. The electric field inhomogeneity sensitivity is an intrinsic characteristic of SF_6 gas insulation systems. In uniform electric fields, discharge propagation to the opposite electrode occurs immediately upon initiation. Conversely, under non-uniform electric field conditions, discharge propagation requires significantly higher applied voltages to satisfy the critical propagation electric field strength threshold. GIS/GIL insulation gas gaps are characterized by a weakly non-uniform electric field distribution, in which the ratio between the maximum and the minimum electric field in the gap is in the range between 2 and 3 [22]. In this paper, the uniformity factor of the electric field (f) of samples is calculated by Equation (1), and the results are shown in Table 1. Clearly, the electric field distribution in samples with plate–plate shielding is more non-uniform than in samples with rod–plate shielding.

$$f = \frac{E_{max}}{E_{av}} = \frac{hE_{max}}{U} \quad (1)$$

where f refers to the uniformity factor of the electric field distribution, 1; E_{max} refers to the maximum electric field strength, kV/mm; E_{av} refers to the average electric field strength, kV/mm; h refers to the height of the gas gap, mm; and U refers to the applied voltage, kV.

Table 1. Sample ID with the corresponding uniformity factor of the surface electric field distribution.

Shielding Electrodes	**Rod–Plate**				**Plate–Plate**			
Surface profile	Gas gap	Concave	Wave	Cylinder	Gas gap	Concave	Wave	Cylinder
Sample ID	(a)	(b)	(c)	(d)	(e)	(f)	(g)	(h)
Location of E_{max}	Electrode surface				Dielectric surface			
f	5.29	3.08	3.18	3.52	1.61	2.82	2.60	1.71

In our previous experiments [24], we employed a photomultiplier tube (PMT) and a high-speed framing camera to observe the spatiotemporal optoelectronic distribution of the discharge process in sample a (f = 5.29) and sample b (f = 2.82) under lightning impulse voltage. Analysis of the discharge images indicates that the discharge inception consistently occurred at the location of the maximum electric field strength. Furthermore, based on the time delay between gap breakdown and the initial detection of the discharge signal, we found that for sample b (f = 2.82), breakdown occurred within 100 ns after discharge inception. This suggests that in slightly non-uniform electric field configurations, the breakdown voltage can be considered representative of the discharge inception voltage. The captured surface flashover images confirm that discharge initiation consistently occurred at the location of the simulated maximum electric field (E_{max}). This observation strongly supports the reliability of the uniformity factor (f) used in this study.

3. Results

3.1. Breakdown Voltage

Figure 4 illustrates the dependence of AC breakdown voltage on gas pressure for the samples, covering a range from 0.1 to 0.4 MPa. In the gas gap with rod–plate shielding, after the introduction of post insulators, E_{max} is still at the electrode surface and its value decreases, as shown in Figure 3a–d, so the breakdown voltage (U_{bd}) of the insulator samples is higher than that of the pure gas gap. However, in the gas gap with plate electrodes, E_{max} shifts from the electrode surface to the insulator surface and its value increases, as shown in Figure 3e–h, so U_{bd} decreases after the introduction of post insulators.

As shown in Figure 5a, in the rod–plate structure, the breakdown voltage increment (ΔU_{bd}) decreases sharply with increasing gas pressure, while ΔU_{bd} only decreases slightly for the case shown in Figure 5b. The most significant distinction between these two sample sets is their electric field non-uniformity: the former exhibits strong electric field non-uniformity (f > 3), while the latter demonstrates moderate non-uniformity (f < 3).

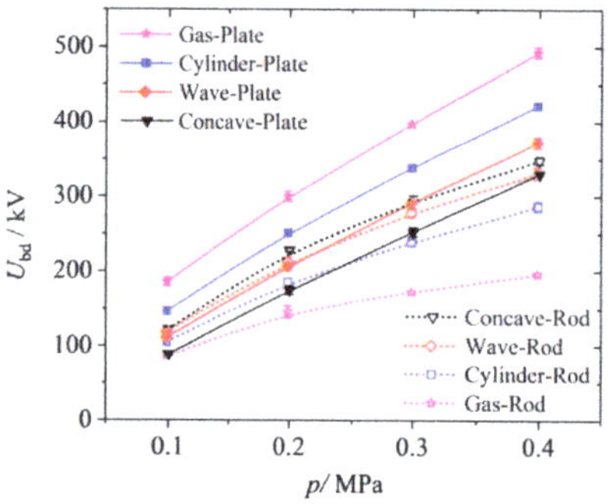

Figure 4. Breakdown voltage of insulator samples as a function of gas pressure.

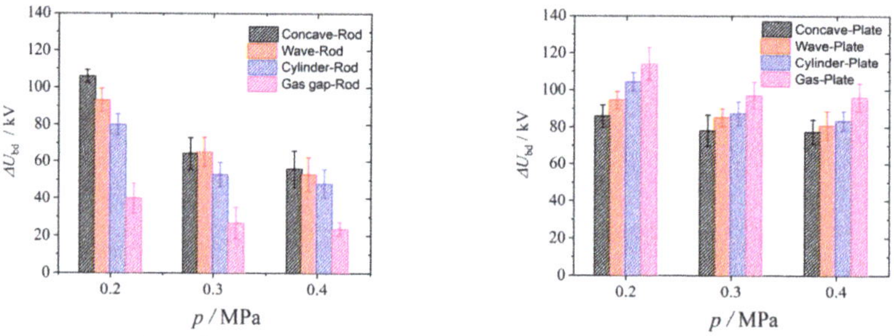

(**a**) Rod–plate. (**b**) Plate–plate.

Figure 5. Increment of the breakdown voltage with the increase in gas pressure.

3.2. Breakdown Electric Field Strength

Theoretically, the discharge inception condition of SF_6 gas ($E_{cr}/p = 88.5$ kV·mm^{-1}·MPa^{-1}) is constant [25,26], which means E_{cr} should increase linearly with gas pressure. Although discharge initiation distorts the background electric field distribution, the AC breakdown voltage in a quasi-uniform electric field gap remains primarily determined by E_{max} in the gas volume [22]. The maximum value of the electric field strength at breakdown voltage (E_{bdmax}) of all samples was calculated by Equation (2), and the results are presented in Figure 6.

$$E_{bdmax} = U_{bd} E_{max} \tag{2}$$

where E_{bdmax} refers to the maximum electric field strength along the insulator surface at breakdown voltage, kV/mm; E_{max} refers to the maximum electric field strength under 1000 V, kV/mm; and U_{bd} refers to the breakdown voltage, kV.

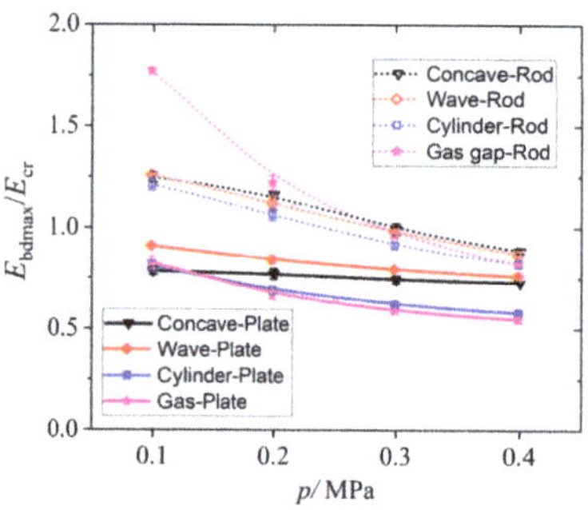

Figure 6. Breakdown electric field strength of insulator samples at different gas pressures.

As shown in Figure 6, the E_{bdmax} of samples under rod–plate shielding is lower than the theoretical value when the gas pressure is higher than 0.3 MPa, while the E_{bdmax} of samples under plate–plate shielding is lower than the theoretical value when the gas pressure is higher than 0.1 MPa. These results may be attributed to two factors: First, since electron swarm studies are typically conducted under low-pressure conditions with small electrode gaps (where the *pd* product is minimal), the established criteria may not be applicable to high-pressure, large-gap configurations. Second, although the samples were polished with sandpaper, protrusions may still remain on the electrode and insulator surface. The breakdown voltage of SF_6 gas gaps exhibits strong sensitivity to the electric field inhomogeneity, particularly at elevated gas pressures [22,27]. Consequently, the measured breakdown voltages are consistently lower than theoretical streamer inception values under such conditions.

3.3. Relationship Between Breakdown Voltage and Maximum Surface Electric Field Strength

Theoretical analysis predicts a significant inverse correlation between U_{bd} and E_{max} in SF_6 gas gaps under identical geometric conditions. However, the descending trend of the U_{bd}-E_{max} curve in this study exhibits a distinct hump; U_{bd} even increases rather than decreasing as E_{max} increases from 0.0718 kV/mm to 0.1015 kV/mm, as shown in Figure 7.

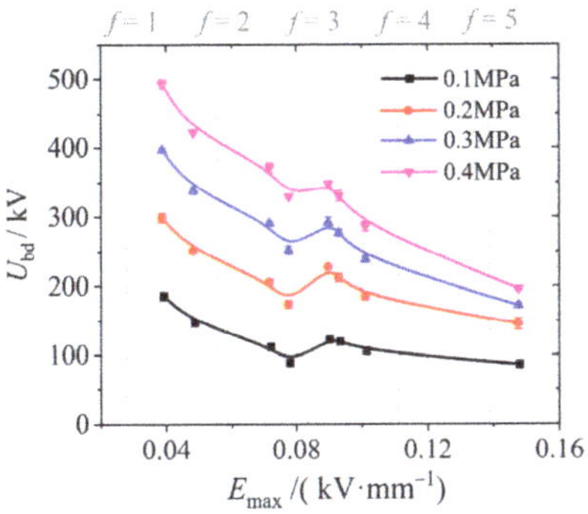

Figure 7. Breakdown voltage of insulator samples as a function of the maximum surface electric field strength.

The samples were divided into two groups according to the discharge initiation position, and Figure 7 was redrawn to obtain the two U_{bd}-E_{max} curves shown in Figure 8. It can be seen that the descending trend of the U_{bd}-E_{max} curve for the DIBD group is faster than that for the EIBD group.

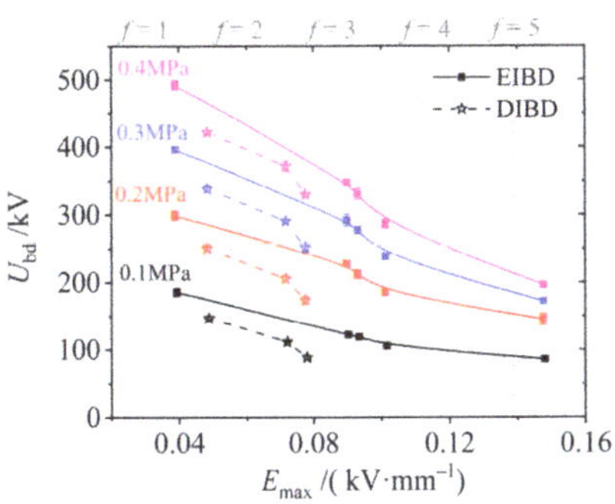

Figure 8. Electric field strength dependence of the breakdown voltage in EIBD and DIBD.

Although the mechanisms of gas discharge inception differ under AC and impulse voltages, experimental results obtained under impulse conditions can still serve as a valuable reference for trends observed in AC voltage scenarios. Here, we utilized our previously acquired dataset comprising breakdown voltages (U_{bd}) under standard lightning impulse and the electric field distribution profiles for 48 distinct post-type insulator configurations. These measurements were obtained in a 36 mm SF_6 gas gap at 0.2 MPa, for which the data points were more densely distributed across the varying range of E_{max} [28]. The breakdown voltage was also divided into two groups: EIBD and DIBD, and the separate phenomenon is more obvious in the U_{bd}-E_{max} curves, as shown in Figure 9. This implies that in experiments with large sample sizes and under standard lightning impulse, the trend of U_{bd} decreasing with the increase in E_{max} in DIBD is more significant.

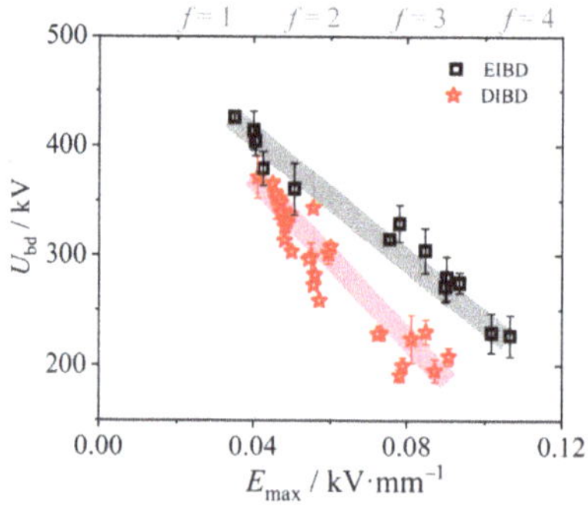

Figure 9. Maximum electric field strength dependence of the breakdown voltage in EIBD and DIBD under a lightning impulse in 0.2 MPa SF_6 gas. The data are partially based on our previous work [28].

4. Discussions

4.1. Influence of Solid Dielectric Surface on Electric Field Distributionn

As illustrated in Figure 3, E-distribution along the dielectric surface exhibits significantly greater complexity compared to that in a gas gap. In a gas medium, the electric field lines originate perpendicularly from the electrode surface and exhibit a monotonic decay with increasing distance from the electrode. Conversely, E-distribution along the surface of a solid dielectric is governed by both the material properties and the geometry of shielding electrodes, resulting in a composite electric field comprising a tangential component parallel to the solid dielectric surface (E_t) and a normal component perpendicular to the solid dielectric surface (E_n). This dual-component characteristic leads to non-monotonic variations in the surface electric field strength between electrodes.

The influence of surface micro-protrusions on the electric field distortion along dielectric interfaces cannot be neglected. To systematically evaluate the influence of microscopic surface protrusions on E-distributions, identical rounded-triangular defects (height: 0.10 mm, base width: 0.20 mm) at the E_{max} locations of both electrode and dielectric surfaces were introduced to samples e and h, respectively. The distorted E-distribution caused by such a micro-protrusion under the breakdown voltage is shown in Figure 10.

Obviously, the enhanced E^*_{max} at the electrode surface is increased by 125%, which is much higher than the 36.9% recorded along the solid dielectric surface. However, E-distribution in a 3 mm long strip-shaped region near the location of E_{max} along the solid dielectric surface significantly increased from 70%E_{cr} to 80%E_{cr}, while only the electric field strength in a circular region with a radius of about 0.2 mm near the protrusion was significantly enhanced. This is also one of the reasons why E_{max} for dielectric surface-initiated discharge is significantly lower than the theoretical value. The electric field simulations are conducted under idealized assumptions, whereas in practice, both electrode and solid dielectric surfaces exhibit dimensional deviations from design specifications and surface roughness. Consequently, for samples with weakly non-uniform electric field configurations ($f < 3$), the experimentally determined E_{bdmax} systematically underestimates theoretical predictions derived from E-distribution. The numerical values obtained from simulation calculations cannot exactly match the actual E-distribution in the samples during breakdown experiments, but they can still accurately characterize the overall E-distribution.

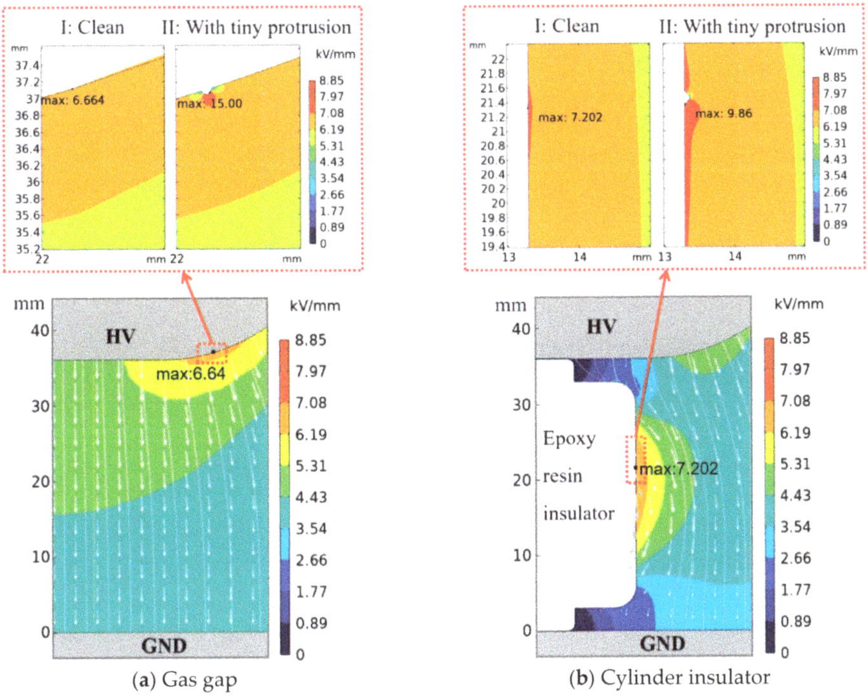

Figure 10. Theoretical electric field distribution at the breakdown voltage of two typical samples under plate–plate electrodes at 0.1 MPa.

4.2. Influence of the Solid Surface on the Critical Conditions for Gaseous Discharge

According to Raethe–Meek's criterion [29–31], the transition from electron avalanche discharge to self-sustained streamer discharge occurs when the following quantitative conditions are met:

$$\int_0^l \alpha_{\text{eff}}(E,p)\mathrm{d}l = 10^8 \tag{3}$$

where α_{eff} is the effective ionization coefficient. Specifically, $\alpha_{\text{eff}} = \alpha - \eta$ if $\alpha > \eta$, and it is zero otherwise. α is the ionization coefficient, 1; η is the attachment coefficient, 1; E refers to the applied electric field strength, kV/mm; p refers to the gas pressure, MPa; and l is the distance that the avalanche can propagate in mm. The integration is normally performed along the electric field line.

The stochastic initiation of gas discharges stems from its nonlinear relationship to the local electric field configuration, where both the spatial extent and geometry of the electric field enhanced regions exhibit a greater determinative influence than E_{max} alone.

The charged particles produced by the discharge will further distort the background E-distribution. After the electron avalanche initiation, the modified spatial E-distribution can be characterized by Equation (3). Streamer propagation becomes sustainable when two critical conditions are met: (1) the electron density at the electron avalanche head exceeds ~10^8, and (2) the avalanche head radius reaches approximately 30 μm [32]. Under these conditions, the space charge-induced electric field becomes sufficient to significantly modify the background E-distribution. Remarkably, streamer advancement can persist even when the enhanced local E_{max} remains slightly below the conventional breakdown threshold (E_a > 75 kV·cm^{-1}·MPa^{-1}). This clearly demonstrates the importance of space charge effects in discharge development [32].

$$E = E_a + E_{sp} = E_a + \frac{N_s e}{4\pi\varepsilon_0 (z_0 + R_s)^2} \tag{4}$$

where E_a is the applied electric field strength, kV/mm; E_{sp} is the electric field strength produced by space charge, kV/mm; N_s is the number of charges at the head of the electron avalanche, 1; R_s is the radius of the head of the electron avalanche, μm; e is the charge of a single electron, C; and z_0 is the axial distance from the start of the secondary electron avalanche, mm.

As depicted in Figure 11, the discharge processes in a pure gas gap are primarily driven by the ionization and attachment of SF_6 gas molecules in the active region, which are highly dependent on the local electric field strength and gas pressure [31–33].

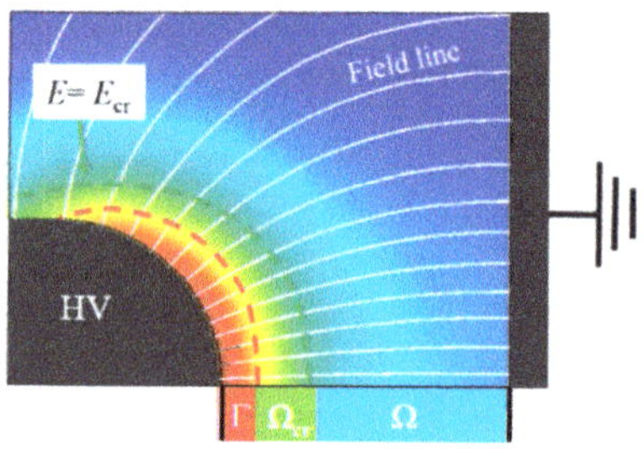

Figure 11. Streamer inception in the gas gap. Generally, there are no electron avalanches in domain ($E < E_{cr}$) as $\alpha_{eff} < 0$. In domain Ω_{cr} ($E > E_{cr}$) the electron avalanches initiate and grow towards the anode. Streamers propagate in the region as the electric field is strong enough.

The traditional gas–solid interface discharge models, as reviewed in [34], generally employ a two-dimensional axisymmetric model with parallel plate electrodes and a cylindrical smooth solid dielectric. Thus, the direction of the applied electric field is parallel to the gas–solid interface.

When E_{max} occurs at the solid dielectric surface, the role of E_n in the discharge process becomes critical, as illustrated in Figure 12. First, the solid dielectric surface acts as a potential barrier, restricting the motion of free electrons and reducing their probability of escaping from the active discharge region. This confinement distorts the local E-distribution, enhancing ionization probability. Second, high-energy photons and electrons bombard the dielectric surface, liberating additional free electrons at the gas–solid interface [11–14]. Under the influence of E_n (perpendicular to the interface), these electrons are driven toward the surface, where they become trapped or bound, leading to surface charge accumulation. Extensive experimental evidence confirms that surface charge accumulation significantly alters the surface flashover voltage and discharge morphology [35–37]. Thus, E_n amplifies the solid surface's positive feedback effect on discharge propagation.

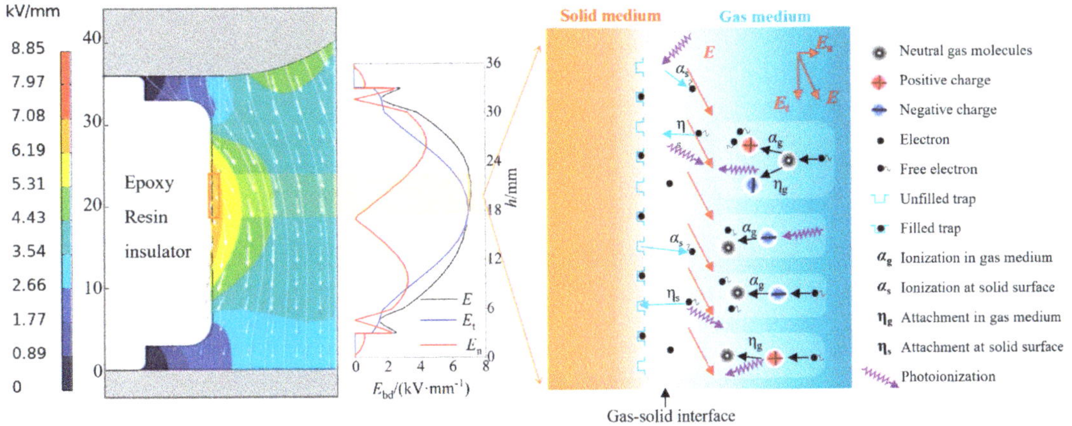

Figure 12. Schematic diagram of the main ionization and adsorption processes at the SF_6–epoxy gas–solid interface. The size and direction of the red arrows near the interface represent the strength and direction of the local electric field, respectively.

The process of generating and trapping free electrons on the solid surface can be simplified to the surface ionization coefficient (α_d) and the surface attachment coefficient (η_d), which are mainly determined by the local electric field strength. Therefore, the effective ionization coefficient α_{eff} in the Raether–Meek criterion for solid dielectric surface-initiated discharge should be modified as follows:

$$\alpha_{eff}(E,p) = \alpha_{g\_eff}(E,p) + \alpha_{s\_eff} = (\alpha_g(E,p) - \eta_g(E,p)) + (\alpha_s(E) - \eta_s(E)) \quad (5)$$

where α_{eff}, α_{g_eff}, and α_{s_eff} represent the effective ionization coefficient of the gas–solid interface, gas gap, and solid dielectrics, 1; α_g and α_s represent the ionization coefficient of SF_6 gas gap and solid dielectrics, 1; and η_g and η_s represent the attachment coefficient of the SF_6 gas gap and solid dielectrics, 1. E refers to the applied electric field strength, kV/mm; p refers to the gas pressure, MPa.

α_{g_eff} varies with gas pressure, while α_{s_eff} is only related to the intrinsic parameters of the solid dielectric material. As the gas pressure increases, the magnitude of E_{cr} required for the initiation of gas–solid interface discharge also rises. As a result, the role of electrons emitted from the solid dielectric surface in the discharge process becomes more significant, potentially generating seed electrons that trigger the initiation of discharge.

In conclusion, both E_n and the gas pressure increment contribute to enhancing the effective ionization process on the gas–solid interface. E_n promotes the ionization of electrons on the solid surface. As gas pressure increases, the applied electric field strength required for discharge initiation rises correspondingly. This pressure-dependent enhancement effect is particularly pronounced for E_n. Combined with factors that distort the electric field, such as defects on the solid surface, the critical condition for gas discharge initiation at the solid dielectric surface is generally lower than that at the electrode surface.

It must be emphasized that the above analysis is qualitatively derived from macroscopic experimental results in conjunction with fundamental discharge theories, and is intended solely as a reference for researchers engaged in the structural design and experimental investigation of related insulation components. More accurate theoretical studies should be based on controlled-variable investigations of surface conditions and precise observations of optoelectronic signals during the discharge inception phase.

5. Conclusions

Although it is widely recognized that introducing a solid dielectric weakens gaseous insulation, the critical conditions governing surface discharge inception—often the immediate trigger for dielectric failure—remain largely unexplored. This paper experimentally measured the AC breakdown voltage of surface-initiated discharge and electrode-initiated discharge in SF_6 gas at pressures ranging from 0.1 MPa to 0.4 MPa. Combined with electrostatic field simulations, the study analyzed the influence of the solid dielectric on the initiation and development of discharge. The conclusions are as follows:

(1) Under AC voltage conditions, the critical electric field for solid dielectric surface-initiated breakdown in SF_6 gas is significantly lower than that of metal electrode surface-initiated breakdown under identical electric field distribution. Although the discharge mechanisms differ, the same disparity phenomenon is also observed under standard lightning impulse voltage.

(2) The discharge process on dielectric surfaces demonstrates higher sensitivity to microscopic irregularities. According to the electrostatic field simulation results, while micro-protrusions significantly amplify the maximum electric field strength on the electrode surface, their influence on the electric field distribution in the adjacent gas region remains limited. In contrast, similar protrusions on a solid dielectric surface cause more extensive distortion of the electric field along the direction of the electric field lines.

(3) From the perspective of discharge mechanisms, a qualitative analysis indicates that as gas pressure rises, the critical electric field strength required to initiate discharge at the gas–solid interface increases, and more free electrons are released from the solid dielectric surface. This is further promoted by the normal component of the electric field, as well as local electric field distortions due to surface defects, resulting in significantly lower discharge initiation thresholds compared to a pure gas gap.

(4) These conclusions underscore the qualitative role of localized electric field distortion and material surface properties in discharge behavior. Further research employing partial discharge imaging and surface functionalization techniques is essential to quantitatively elucidate the proposed mechanisms and generalize these findings.

Author Contributions: Methodology, L.L., Q.Z. and X.L.; validation, Q.Z. and X.L.; formal analysis, L.L. and X.P.; investigation, L.L., Q.Z. and X.L.; resources, Q.Z.; data curation, L.L. and X.L.; writing—original draft preparation, L.L., X.P. and Z.W.; writing—review and editing, L.L. and S.Y.; visualization, Z.W. and S.Y.; supervision, Q.Z. and X.L.; funding acquisition, L.L. All authors have read and agreed to the published version of the manuscript.

Funding: This work was supported by the National Natural Science Foundation of China (grant number 52307174).

Data Availability Statement: Data is contained within the article.

Conflicts of Interest: Lin Liu, Xiangyang Peng, Zheng Wang, and Shihu Yu are employed by the company Electric Power Research Institute of Guangdong Power Grid Co., Ltd. The remaining authors declare that the research was conducted in the absence of any commercial or financial relationships that could be construed as a potential conflict of interest.

References

1. Koch, H. *Gas Insulated Substations (GIS)*; Wiley-IEEE Press: Hoboken, NJ, USA, 2014; pp. 34–40.
2. Li, B. *SF6 High Voltage Apparatus*, 6th ed.; China Machine Press: Ningbo, China, 2019; pp. 79–96.
3. Han, X.C.; Sun, X.; Chen, H.B.; Qiu, N.; Lyu, D.; Wang, N.H.; Wang, X.N.; Zhang, J.L. The overview of development of UHV AC transmission technology in China. *Proc. CSEE* **2020**, *40*, 4237–4386. [CrossRef]
4. Küchler, A. *High Voltage Engineering*; Springer: Berlin/Heidelberg, Germany, 2018; pp. 159–215.

5. Niu, S.Y.; Zhao, Q.Y.; Niu, S.X.; Jian, L.N. A Comprehensive investigation of thermal risks in wireless ev chargers considering spatial misalignment from a dynamic perspective. *IEEE J. Emerg. Sel. Top. Ind. Electron.* **2024**, *5*, 1560–1571. [CrossRef]
6. Nitta, T.; Shibuya, Y.; Fujiwara, Y.; Arahata, Y.; Takahashi, H.; Kuwahara, H. Factors controlling surface flashover in SF6 gas insulated systems. *IEEE Trans. Power Appar. Syst.* **1978**, *PAS-97*, 959–968. [CrossRef]
7. Laghari, J. Spacer flashover in compressed gases. *Trans. Electr. Insul.* **1985**, *20*, 83–92. [CrossRef]
8. Sudarshan, T.; Dougal, R. Mechanisms of surface flashover along solid dielectrics in compressed gases: A review. *IEEE Trans. Electr. Insul.* **1986**, *21*, 727–746. [CrossRef]
9. Xing, Y.Q.; Wang, Z.W.; Liu, L.; Xu, Y.; Yang, Y.; Liu, S.; Zhou, F.S.; He, S.; Li, C.Y. Defects and failure types of solid insulation in gas-insulated switchgear: In situ study and case analysis. *High Volt.* **2022**, *7*, 158–164. [CrossRef]
10. Li, X.; Liu, W.D.; Xu, Y.; Chen, W.J.; Bi, J.G. Surface charge accumulation and pre-flashover characteristics induced by metal particles on the insulator surfaces of 1100 kV GILs under AC voltage. *High Volt.* **2020**, *5*, 134–142. [CrossRef]
11. Verhaart, H.; Verhage, A. Insulator flashover in SF6 gas. *Kema Sci. Technol. Rep.* **1988**, *6*, 179–228.
12. Jorgenson, R.E.; Warne, L.K.; Neuber, A.A.; Krile, J.; Dickens, J.; Krompholz, H.G. *Effect of Dielectric Photoemission on Surface Breakdown an LRRD Report*; Sandia National Laboratories: Livermore, CA, USA, 2003; p. 79.
13. Chvyreva, A. *Creeping Sparks: A Study on Surface Discharge Development-Electrodeless*; Eindhoven University of Technology: Eindhoven, The Netherlands, 2016; pp. 40–45.
14. Chvyreva, A.; Pancheshnyi, S.; Christen, T.; Pemen, A.J.M. Raether–Meek criterion for prediction of electrodeless discharge inception on a dielectric surface in different gases. *J. Phys. D Appl. Phys.* **2018**, *51*, 115202. [CrossRef]
15. *IEC 60296:2020*; Fluids for Electrotechnical Applications—Mineral Insulating Oils for Electrical Equipment. International Electrotechnical Commission: Geneva, Switzerland, 2020.
16. *IEC 60243-1:2013*; Electric Strength of Insulating Materials—Test Methods—Part 1: Tests at Power Frequencies. International Electrotechnical Commission: Geneva, Switzerland, 2013.
17. Dawson, G.; Winn, W. A model for streamer propagation. *Z. Phys. A Hadron. Nucl.* **1965**, *183*, 159–171. [CrossRef]
18. Gallimberti, I.; Wiegart, N. Streamer and leader formation in SF_6 and SF_6 mixtures under positive impulse conditions. I. Corona development. *J. Phys. D Appl. Phys.* **1986**, *19*, 2351–2361. [CrossRef]
19. Gallimberti, I.; Wiegart, N. Streamer and leader formation in SF_6 and SF_6 mixtures under positive impulse conditions. II. Streamer to leader transition. *J. Phys. D Appl. Phys.* **1986**, *19*, 2363–2379. [CrossRef]
20. Niemeyer, L.; Ullrich, L.; Wiegart, N. The mechanism of leader breakdown in electronegative gases. *IEEE Trans. Electr. Insul.* **1989**, *24*, 309–324. [CrossRef]
21. Hayakawa, N.; Hatta, K.; Okabe, S.; Okubo, H. Streamer and leader discharge propagation characteristics leading to breakdown in electronegative gases. *IEEE Trans. Dielectr. Electr. Insul.* **2006**, *13*, 842–849. [CrossRef]
22. Seeger, M.; Niemeyer, L.; Bujotzek, M. Partial discharges and breakdown at protrusions in uniform background fields in SF_6. *J. Phys. D Appl. Phys.* **2008**, *41*, 185204. [CrossRef]
23. Seeger, M.; Niemeyer, L.; Bujotzek, M. Leader propagation in uniform background fields in SF_6. *J. Phys. D Appl. Phys.* **2009**, *42*, 185205. [CrossRef]
24. Liu, L.; Li, X.A.; Wen, T.; Bo, Q.H.; Ma, J.T.; Zhang, Q.G. Criteria and propagation processes of electrode initiated and electrodeless-initiated discharge in SF_6. *Phys. Plasmas* **2019**, *26*, 023512. [CrossRef]
25. Pedersen, A. Criteria for spark breakdown in sulfur hexafluoride. *IEEE Trans. Power Appar. Syst.* **1970**, *89*, 2043–2048. [CrossRef]
26. Govinda, R.; Dincer, M. Measurement of ionization and attachment coefficients in SF_6 and SF_6+N_2. *J. Phys. D Appl. Phys.* **1982**, *53*, 8562–8567. [CrossRef]
27. Zhang, Q.G.; Li, X.A.; Liu, L.; Wen, T.; Zhao, J.P. "One-Zero" phenomenon of discharge induced by defects in SF6 insulation systems in GIS. *High Volt. Eng.* **2019**, *45*, 2689–2698. [CrossRef]
28. Liu, L.; Li, X.A.; Wen, T.; Zhang, R.; Wu, Z.C.; Zhao, J.P.; Zhang, Q.G. Investigations on the surface electric field distribution features related to insulator flashover in SF_6 gas. *IEEE Trans. Dielectr. Electr. Insul.* **2019**, *26*, 1588–1595. [CrossRef]
29. Loeb, L. Fundamental processes of electrical discharge in gases. *Nature* **1940**, *146*, 729–730. [CrossRef]
30. Meek, J. A theory of spark discharge. *Phys. Rev.* **1940**, *57*, 722–728. [CrossRef]
31. Raether, H. *Electron Avalanches and Breakdown in Gases*; Butterworths: Oxford, UK, 1964; pp. 15–22.
32. Raju, G. *Dielectrics in Electric Fields: Tables, Atoms, and Molecules*, 2nd ed.; CRC Press: Boca Raton, FL, USA, 2017; pp. 383–478.
33. Raju, G.; Liu, J.F. Simulation of electrical discharges in gases. Nonuniform electric fields. *IEEE Trans. Dielectr. Electr. Insul.* **1995**, *2*, 1016–1041. [CrossRef]
34. Li, Z.; Liu, J.; Ohki, Y.; Chen, G.; Li, S.T. Surface flashoverin 50 years: II. Material modification, structureoptimisation, and characteristics enhancement. *High Volt.* **2025**, *10*, 243–278. [CrossRef]
35. Deng, J.B.; Matsuoka, S.; Kumada, A.; Hidaka, K. The influence of residual charge on surface discharge propagation. *J. Phys. D Appl. Phys.* **2010**, *43*, 495203. [CrossRef]

36. Zhang, Q.G.; Wang, Z.F.; Gu, W.G.; Qiu, Y.C. Effect of surface charges on impulse flashover voltages of spacer in SF6. In Proceedings of the 1998 International Symposium on Electrical Insulating Materials, 1998 Asian International Conference on Dielectrics and Electrical Insulation, 30th Symposium on Electrical Insulating Materials, Toyohashi, Japan, 30 September 1998; pp. 445–448. [CrossRef]
37. Li, C.Y.; Hu, J.; Lin, C.J.; Zhang, B.Y.; Zhang, G.X.; He, J.L. Surface charge migration and DC surface flashover of surface-modified epoxy-based insulators. *J. Phys. D Appl. Phys.* **2017**, *50*, 065301. [CrossRef]

energies

MDPI

Review

Full-Life-Cycle Management of High-Voltage Bushings Based on Digital Twin: Typical Scenarios, Core Technologies, and Research Prospects

Weiwei Chi [1], Tao Wang [1], Jichao Zhang [1], Zili Wang [2] and Chuyan Zhang [2,*]

1 State Grid Hebei Electric Power Co., Ltd. Xiong'an New Area Power Supply Company, Xiong'an 071001, China; xa_chiww@he.sgcc.com.cn (W.C.); xa_wangt@he.sgcc.com.cn (T.W.); xa_zhangjc@he.sgcc.com.cn (J.Z.)

2 School of Artificial Intelligence, China University of Geosciences Beijing, Beijing 100083, China; wzl@email.cugb.edu.cn

* Correspondence: zcy@cugb.edu.cn

Abstract: High-voltage (HV) bushings are critical hub components in power systems, whose operational reliability is paramount to the safety and stability of transmission and distribution infrastructure. Conventional management paradigms are hampered by challenges such as information silos, reactive maintenance, and imprecise condition assessment, rendering them in-adequate for the evolving demands of modern power systems. Digital twin technology, by creating a high-fidelity, re-al-time interplay between physical entities and their virtual counterparts, provides a revolutionary pathway toward the intelligent full-life-cycle management (FLCM) of HV bushings. This paper presents a review of the current state of research in this domain. It begins by reviewing research on the construction a five-dimensional digital twin framework that encompasses the entire lifecycle: design, manufacturing, operation and maintenance (O&M), and decommissioning. Subsequently, it delves into the application paradigms of digital twins across typical scenarios, including external insulation design, intelligent condition assessment, insulation defect identification, fault diagnosis, and predictive maintenance. The paper then examines the core technological underpinnings, such as multi-physics coupled modeling, multi-source heterogeneous data fusion, and data-driven model updating and condition assessment. Finally, it identifies current challenges related to data, models, standards, and costs, and offers a forward-looking perspective on future research directions, including group digital twins, deep integration with artificial intelligence, edge-side deployment, and standardization initiatives. This work aims to provide a theoretical reference and technical guidance for advancing the intelligent O&M of HV bushings and bolstering grid security.

Keywords: HV bushing; digital twin; intelligent operation and maintenance; insulation defect; multi-physics simulation; machine learning

1. Introduction

In response to the profound shifts in the global energy sector and the pursuit of "carbon peaking and carbon neutrality" objectives of China [1,2], the establishment of a new power system centered around renewable energy is rapidly gaining momentum. Amidst this transformation, high-voltage direct current (HVDC) transmission stands out as a pivotal technology for grid interconnection due to its unparalleled capabilities in transmitting large capacities over extended distances [3,4]. HV bushings are indispensable and critical components in modern power systems, serving as the primary insulation and feed-through devices for conductors at earthed barriers such as transformer tanks, circuit breaker walls, and substation enclosures. The operational reliability of HV bushings is directly linked to the stability and security of the entire electrical grid [5]. A catastrophic bushing failure can result in explosive equipment damage, widespread power outages, and significant safety hazards to personnel. Consequently, the implementation of effective condition-based maintenance (CBM) and FLCM strategies for HV bushings is of paramount importance. Despite their critical role, the FLCM of HV bushings is fraught with multifaceted challenges, stemming from their complex structure, harsh operating environment, and the insidious nature of their degradation mechanisms.

Modern HV bushings, particularly resin-impregnated paper (RIP) [6] or resin-impregnated synthetic (RIS) [7] types, exhibit intricate multi-layer structures. These structures comprise concentric layers of conductive foils and insulating materials, creating a complex capacitive grading system. This internal complexity makes it extremely difficult to directly visualize or model internal defects, such as partial discharge (PD) inception, moisture ingress, or delamination. The thermo-mechanical stresses arising from load currents and ambient temperature variations further exacerbate material aging, leading to a non-linear and coupled degradation process that is challenging to predict.

The degradation of HV bushing insulation is typically a slow, progressive process. Early-stage faults, such as the development of micro-cavities or slight moisture contamination, often produce only subtle changes in macroscopic diagnostic parameters. By the time these changes become apparent through conventional monitoring, the insulation may have already suffered irreversible damage. This "silent" progression from a healthy state to failure creates a significant blind spot in maintenance planning, increasing the risk of unexpected outages.

Moreover, HV bushings are continuously exposed to a combination of electrical, thermal, mechanical, and environmental stressors. These include fluctuating voltage levels, variable load currents, extreme weather conditions (temperature, humidity, pollution), and mechanical vibrations. The synergistic effect of these stressors accelerates aging and complicates the task of isolating the root cause of an observed anomaly, making a definitive diagnosis challenging.

To address these challenges, a variety of online and offline monitoring techniques have been developed. However, each method possesses inherent limitations that hinder the realization of true FLCM. The conventional strategies of regular inspections, periodic preventive testing, and post-disaster analysis represent a passive, responsive management model [8]. This approach is fraught with deficiencies: it creates information silos, as data from design, manufacturing, and operations remain disconnected, precluding a comprehensive lifecycle perspective; it lacks effective condition assessment tools, leading to either over-maintenance or under-maintenance; and it offers no predictive insights into potential failures, hindering a shift from reactive to proactive risk management. These shortcomings fall short of the intelligent, streamlined management demands of the emerging power system.

Traditional periodic maintenance, involving offline tests like dielectric loss factor ($\tan\delta$) measurement, capacitance measurement, and PD detection at power frequency, has been the industry standard for decades. These tests provide only a snapshot of the bushing's condition at a specific time, missing dynamic changes and transient faults that occur between testing intervals. Furthermore, offline testing requires scheduled outages, which is economically undesirable and operationally disruptive. The results can be significantly influenced by temperature and humidity, requiring complex correction and introducing potential for misinterpretation.

Online monitoring systems have been widely deployed to overcome the limitations of offline testing by continuously tracking key parameters. However, different monitoring systems often operate in silos, providing fragmented data streams. It is difficult to synthesize this multi-source data to form a holistic understanding of the bushing's health. For instance, a slight increase in tan δ might be insignificant on its own but critical when correlated with a rising temperature trend and intermittent PD pulses. Most online systems are based on pre-set thresholds. They are proficient at detecting that a problem exists but are often incapable of pinpointing the fault's location, type, and root cause. For example, PD localization within a bushing remains a significant technical challenge. These systems are primarily reactive or, at best, diagnostic. They lack the capability to accurately predict the remaining useful life (RUL) of the bushing, which is the cornerstone of proactive CBM and asset management planning.

Recent advancements in digital twin technology present a transformative solution to these challenges. By creating a virtual model that mirrors and interacts with its physical counterpart in real-time, digital twins enable comprehensive monitoring, diagnostic forecasting, and decision optimization for complex equipment throughout their life cycles [9–11]. The integration of digital twin technology into the management of high-voltage bushings promises to bridge information gaps, facilitate a holistic analysis from micro-level material aging to macro-level system performance, and steer the operational model towards a more intelligent, predictive approach. While digital twins have been explored in power equipment like transformers and circuit breakers [12], a systematic framework encompassing the entire life cycle of high-voltage bushings, along with an in-depth examination of its application scenarios and core technologies, still requires greater emphasis and research attention.

The foundational research in this area focuses on establishing the architectural framework for a bushing digital twin. A prominent trend is the move beyond a simple three-dimensional (3D) geometric model towards a multi-dimensional framework. For instance, researchers have proposed a five-dimensional (5D) digital twin model [13]. A core component of a high-fidelity DT is the physics-based model. Research in this area concentrates on developing accurate multi-physics models that can simulate the complex internal behavior of bushings. These models simulate the temperature distribution within the bushing under varying load and ambient conditions, which is crucial for understanding thermal aging.

While physics models provide a fundamental understanding, data-driven approaches are essential for handling complex, non-linear relationships and making accurate predictions from real-world monitoring data. Machine learning and deep learning algorithms are being trained on historical data to automatically classify different fault types. Researchers are exploring various algorithms, including Recurrent Neural Networks (RNNs) and Long Short-Term Memory (LSTM) networks, to model the long-term degradation trajectory of bushings and predict their RUL. These models often integrate multi-source data to improve prediction accuracy. A key research direction is the fusion of physics-based and data-driven models to create a hybrid digital twin. For example, physics-based simulation data can be used to augment training datasets, while data-driven models can be used to calibrate and update the parameters of physics models, creating a self-evolving and more accurate twin.

Current literature demonstrates the application of bushing digital twins in several key scenarios, as illustrated in Table 1. These scenarios primarily include: condition assessment, fault diagnosis and localization, and life prediction and maintenance decision-making. Specifically, condition assessment provides a comprehensive health index by fusing multi-dimensional data, offering a more nuanced view than single-parameter thresholds. Fault diagnosis integrates PD data with electromagnetic simulation models to pinpoint the 3D location of internal discharges, representing a significant advancement over traditional PD detection. Life prediction simulates future degradation scenarios under different operating conditions to predict the RUL and optimize maintenance schedules.

Table 1. The application of bushing digital twins in several key scenarios.

Application Scenarios	Core Technology/Framework	Recommended Method	Typical Related Literature
Condition assessment	Physical model + Data-driven model	Multi-sensor integration, multi-physics field coupling, condition evaluation	[14–25]
Fault diagnosis and localization	Physical model + Data-driven model	Multi-physics field coupling, data fusion, high-performance computing	[15–17,26–39]
Life prediction and maintenance decision-making	Five-dimensional model	Data fusion, high-performance computing, visualization	[9–13,40–46]

This paper provides a comprehensive review and future outlook on the life cycle management of high-voltage bushings using digital twins. The key contributions include: emphasizing a five-dimensional management framework that integrates design, manufacturing, operation, and retirement stages, offering a strategic design for bushing management; a detailed exploration of digital twin applications in design optimization, intelligent condition assessment, fault diagnosis, and predictive maintenance; an organized overview of the underlying technologies, such as multi-physics modeling, data fusion, and intelligent algorithms; and a summary of current research challenges and a prospective view on future developments. The findings aim to provide theoretical and technical guidance for enhancing the intelligent operation and maintenance of high-voltage bushings, thereby ensuring the security and stability of the power grid.

2. The High-Voltage Bushing Digital Twin Lifecycle Management Framework

The five-dimensional digital twin model serves as the core theoretical framework for constructing a digital twin system, with a consensus reached among multiple studies. These dimensions include the physical entity, virtual model, digital twin data, functionalities (service systems), and connection (interaction). In Reference [40], the researchers identified that data perception and fusion, mechanism-data-knowledge hybrid-driven modeling, virtual reality synchronization, and dynamic optimization and collaborative control for multi-level parameters are the key technologies for constructing digital twin models. Furthermore, Tao et al. were the first to explicitly define the aforementioned five-dimensional model of digital twins [41,42] and to elaborate on its technical roadmap for industrial applications [43]. These concepts have garnered support from other scholars [44], not only in terms of their definition, characteristics, applications, and design, but also in their foundational principles. For high-voltage bushings, the logical connections between these dimensions are illustrated in Figure 1.

In the lifecycle management of the high-voltage bushing digital twin, the physical entity is clearly defined as the high-voltage bushing and its components, including specialized sensing systems for the bushing. The virtual model offers a detailed digital representation of the high-voltage bushing, capable of simulating the structure, performance, state, and even the evolution patterns of the physical entity. Digital twin data serve as the foundation for driving the model and are core to supporting intelligent decision-making, encompassing design and manufacturing data, operational state data, multi-physics simulation data, and fault experimental data. Functionalities refer to the various applications provided by the digital twin system, such as condition monitoring, fault diagnosis, lifespan prediction, and optimization decision-making for the bushing, representing the ultimate value of the digital twin. The connection is the information exchange link between the physical entity and the virtual model, ensuring real-time, two-way, and reliable data transmission, typically utilizing IoT (Internet of Things) communication technologies.

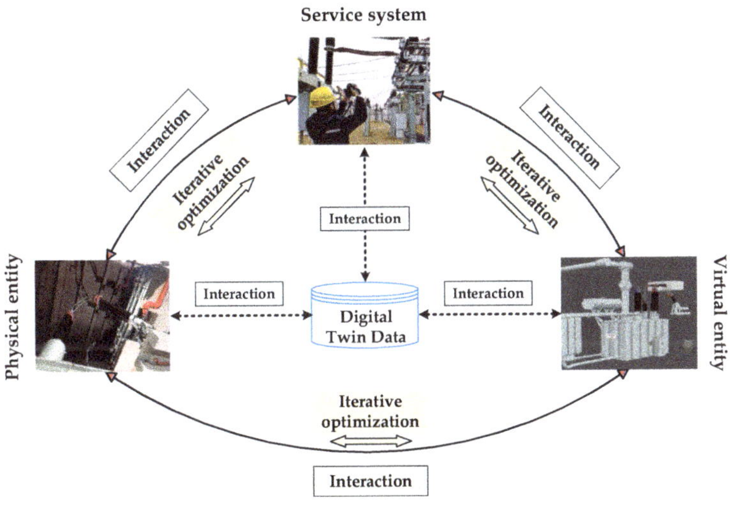

Figure 1. The five-dimensional model serves as the HV bushing digital twin.

This five-dimensional model provides a structured theoretical support for the lifecycle management of the high-voltage bushing digital twin, enabling intelligent closed-loop lifecycle management of the bushing from design to retirement. During the design phase, the virtual model can conduct multi-physics simulation optimization, record design parameters as part of the digital twin data, and support scheme comparison and selection through functionality applications, ultimately achieving design-manufacturing collaboration. In the manufacturing phase of the bushing product, manufacturing process data can be feedback to the virtual model in real-time, used for quality traceability and future process optimization in later stages. In the online operation and maintenance phase of the bushing, operational data of the physical entity are synchronized to the virtual model in real-time through the connection, while functionalities provide condition assessment, fault diagnosis, and predictive maintenance. Finally, in the retirement phase of the high-voltage bushing, lifespan evaluation and retirement decisions are supported based on the accumulated digital twin data, and functionality applications assist in the green recycling and reuse of equipment. Thus, the five-dimensional model tightly connects each link from the "birth" to the "death" of the high-voltage bushing, realizing a data-driven closed-loop lifecycle management.

This management technology route integrates data from multiple sources and formats, ensuring the real-time flow of data between the physical and digital realms, breaking information barriers, which is challenging for traditional management models. In terms of model construction, the digital twin will also surpass traditional single static models, fully leveraging the role of multi-physics (electric field, thermal field, force field, etc.) coupling, realizing state mapping from the macro to the micro level. Based on digital twin data and the virtual model, the high-voltage bushing can transition from "passive maintenance" to "active operation and maintenance," supporting diverse intelligent services and significantly enhancing the operational reliability of the high-voltage bushing. This is an optimal theoretical framework for constructing a new paradigm of asset management for high-voltage power equipment in the context of building a new power system.

3. Typical Application Scenarios of Digital Twins in High-Voltage Bushing Management

Based on the aforementioned five-dimensional theoretical framework for the lifecycle management of high-voltage bushing digital twins, the main application scenarios that can be realized include: design and optimization of external insulation using digital twin simulation, data-driven intelligent condition assessment, identification of insulation defects through multi-physics field integration, fault diagnosis and root cause analysis through the combination of virtual and physical methods, and predictive maintenance based on trend prediction. To support the digital and intelligent operation and maintenance (O&M) of power equipment, Figure 2 demonstrates the typical logical architecture of applications for a HV bushing digital twin.

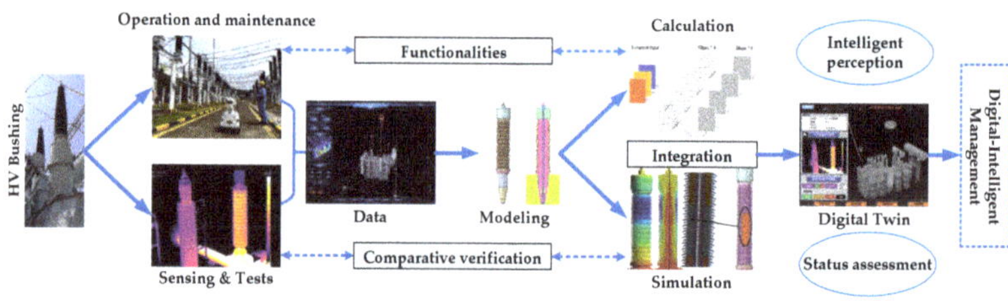

Figure 2. Typical application logic of a HV bushing digital twin.

3.1. Design and Optimization of External Insulation Using Digital Twin Simulation

The insulation of a high-voltage bushing consists of two parts: external insulation and internal insulation. External insulation primarily achieves electrical isolation between the connection end of the internal conductor core and the external busbar, and the grounding shell or wall of transformers and other equipment. Its design and optimization are similar to those of insulator sheds. Internal insulation, on the other hand, achieves electrical isolation between the internal conductor core and the external insulation shed, facing similar issues as the winding insulation of transformers or motors.

The performance of the external insulation of high-voltage bushings, especially its pollution flashover and rain flashover characteristics in complex environments, is a core factor determining its operational reliability [47–49]. Traditional external insulation design mainly relies on empirical formulas, standard tests, and simplified electric field models, which are difficult to accurately simulate the complex interactions between variable environmental conditions and insulation structures during actual operation. This design approach often leads to excessive design margins, resulting in material waste and increased costs, or insufficient design margins. Digital twin technology provides a revolutionary virtual testing and optimization framework for external insulation design by constructing high-fidelity multi-physics field coupling models.

The design optimization process (shown in Figure 3) for external insulation using digital twins begins with the construction of a virtual model that includes the geometric structure of the bushing, material properties, and surrounding environmental parameters. The core of this model is the multi-physics field coupling simulation of electric field, flow field, and particle field. By setting different environmental parameters (such as equivalent salt density, non-soluble deposit density, rainfall rate, wind speed, etc.) and structural parameters (such as shed shape, shed diameter, shed protrusion, rod diameter, etc.) in the virtual environment, designers can conduct thousands of virtual tests without the need to manufacture expensive physical prototypes. The twin model outputs key performance indicators, such as maximum electric field strength and flashover probability. Combined with optimization algorithms (such as genetic algorithms—GA, particle swarm optimization—PSO), the system can automatically find the most material-efficient and structurally optimized design solutions that meet specific electrical performance requirements. This not only significantly shortens the research and development cycle and reduces costs but more importantly, it achieves a shift from "meeting standards" to "performance optimization," providing high-reliability equipment assurance for extra-high voltage projects in extreme environments.

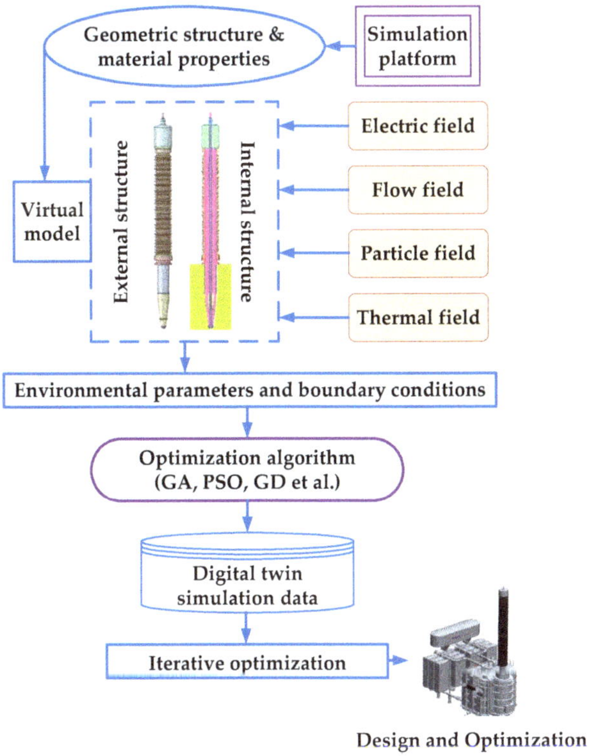

Figure 3. Design and optimization process of external insulation using digital twin simulation.

Research has detailed numerical simulation methods for the pollution accumulation and flashover process of insulators based on multi-physics (electric field, flow field, thermal field), which can provide a technical foundation for constructing digital twin models for high-voltage bushing external insulation. The application of these technologies is exemplified in the packaging of high-voltage, high-power modules, where insulation performance is a primary concern [50]. Here, multiphysics coupling simulations—encompassing electrical, thermal, and mechanical stresses—enable the analysis of how structural parameters affect physical field distributions, which in turn provides critical parameters for insulation design. In another study, researchers leveraged multiphysics simulations to design the external insulation for a 1100 kV GIS SF6 gas-insulated composite bushing [14]. Similarly, this approach facilitates the integration of real-time inversion calculations and simulations of transient temperature and electric fields into the digital twins of HVDC bushings [15] and high-voltage switchgear [26]. The significance of this technology is further underscored by its vital role in the defect analysis of transformer main insulation [27].

Based on these foundations, high-fidelity multi-physics field coupling digital twin models of high-voltage bushings can be constructed using simulation software such as COMSOL Multiphysics [28,29]. The electrical properties of external insulation materials and dielectrics are input into the model, considering the impact of complex environmental factors on breakdown field strength. Then, the external insulation parameters of the high-voltage bushing are set as parametric variables of the model, and the optimization objective is defined as: under given pollution and humidity conditions, the maximum surface electric field strength is below the critical corona inception value. Subsequently, the built-in optimization algorithm module can be used to automatically iterate the simulation calculations, ultimately obtaining the optimized structural solution for the external insulation of the high-voltage bushing. Other research has collected a large number of manually polluted external insulation test results as a dataset, using neural network models to predict pollution flashover voltage, thereby analyzing the impact of external insulation structural parameters on flashover and achieving the optimization of the structure [30].

Furthermore, modeling costs can be reduced through the application of dimensionality reduction algorithms. The computational time for conventional multiphysics simulation methods typically ranges from several hours to even days. In contrast, some studies have employed dimensionality reduction techniques to propose a data-driven, real-time simulation model for the temperature distribution of UVDC transformer bushings [16], which can shorten the model's computation time to the second-level.

3.2. Data-Driven Intelligent Condition Evaluation

The condition evaluation of high-voltage bushings is the core basis for operational and maintenance decisions. Traditional methods mainly rely on periodic offline inspections (such as dissolved gas analysis—DGA, dielectric loss measurement—tan δ) and online monitoring of single or a few parameters (such as top oil temperature) [17,31,51]. This approach has two major limitations: first, the data dimension is single, making it difficult to fully reflect the complex health status of the bushing under the coupling of electrical, thermal, and mechanical stresses; second, the assessment thresholds are mostly fixed values, which cannot adapt to dynamic conditions such as load changes and environmental fluctuations, easily leading to misjudgment or missed judgment. Digital twins, by integrating multi-source heterogeneous data, build data-driven dynamic assessment models, achieving a leap from "point monitoring" to "holistic profiling."

Under the digital twin framework, a comprehensive intelligent condition evaluation system first collects multi-dimensional data from the physical entity, including: dissolved gas analysis (DGA) data, partial discharge (PD) data, dielectric loss factor (tan δ), capacitance value, temperatures at the top, middle, and bottom of the bushing, as well as operating condition data such as ambient temperature, humidity, and load current. The twin data platform cleans, aligns, and spatiotemporally correlates these data. Subsequently, the behavioral model in the virtual model uses machine learning or deep learning algorithms (such as Long Short-Term Memory networks—LSTM [32], Convolutional Neural Networks—CNN [52], or ensemble learning models like Random Forest [18]) to learn the deep feature mapping relationships of the bushing's evolution from normal to faulty conditions from massive historical data.

The core advantage of this assessment model lies in its dynamic and comprehensive characteristics. It does not view any parameter in isolation but treats all parameters and their associated operating condition information as a whole input, outputting a comprehensive health index that dynamically reflects the health level of the bushing. For example, when the load current increases, the model automatically adjusts the tolerance threshold for temperature rise; when the ambient humidity increases, the model recalibrates the interpretation of partial discharge signals. By running this assessment model in real-time within the twin model, operation and maintenance personnel can intuitively see the real-time score, trend, and key influencing factors of the bushing's health status, thus achieving a shift from "regular check-ups" to "real-time monitoring," providing precise and quantitative decision-making basis for condition-based maintenance.

3.3. *Identification of Insulation Defects Through Multi-Physics Field Fusion*

Internal insulation defects in high-voltage bushings (such as partial discharge [53], insulation aging, poor conductor connections, etc.) are the primary sources of sudden failures. The signals generated by these defects in the early stages (such as partial discharge pulses, abnormal temperature rises) are very weak and easily affected by on-site electromagnetic interference, making it difficult for traditional single monitoring methods to effectively identify and accurately locate them. Digital twins, by deeply integrating monitoring data with high-fidelity multi-physics simulation models, provide an "X-ray vision" for defect identification, achieving a leap from "signal detection" to "defect localization."

The core of the insulation defect identification paradigm based on digital twins lies in establishing a virtual model that can accurately describe the internal physical laws of the bushing. This model should at least include an electric field model and a thermal field model [19,20,33]. When the online monitoring system captures questionable signals (such as a unique partial discharge pulse or an abnormal temperature rise point), the signal is input into the twin model. The system then initiates an inverse problem solving or pattern recognition process.

3.4. *Virtual-Physical Integrated Fault Diagnosis and Root Cause Analysis*

When a high-voltage bushing fails or a protective action is triggered, rapidly and accurately diagnosing the type of fault and analyzing its root cause is crucial for preventing the escalation of accidents, formulating repair strategies, and avoiding the recurrence of similar incidents. Traditional fault diagnosis relies on post-failure tests, human experience, and the interpretation of limited recorded data, a process that is time-consuming and may have subjective biases. Digital twins, by constructing an interactive and traceable virtual replay platform, achieve a "digital replay" of the fault process, providing a powerful tool for precise root cause analysis [15,34].

The virtual-physical integrated fault diagnosis process begins at the time of fault occurrence [54,55]. The protective devices and monitoring systems on the physical entity record critical data sequences before and after the fault, such as voltage and current waveforms, sudden changes in DGA data, and pressure relief valve actuation signals. These data are immediately uploaded to the twin data platform and serve as inputs and boundary conditions for the virtual model. Operation and maintenance personnel can replay and reproduce the fault process on the virtual model. For example, by inputting the fault current waveform into the twin's electro-thermal-mechanical coupling model, the dynamic evolution of the internal electric field, temperature distribution, and mechanical stress of the bushing can be observed. If the simulation results from the model show that an extreme electric field concentration or thermal stress buildup has occurred at a certain location, exceeding the material's tolerance limit, then that location is highly likely to be the starting point of the fault. By comparing the simulation results under different fault hypotheses (such as insulation breakdown, seal failure, lead short circuit, etc.) with the actual monitoring data, possibilities can be sequentially eliminated, ultimately identifying the most likely fault mode. This "hypothesis-simulation-verification" cyclic analysis makes root cause analysis no longer a "black box" guess, but a logical reasoning based on physical laws, greatly enhancing the depth and credibility of the analysis.

3.5. Predictive Maintenance Based on Trend Prediction

The core of predictive maintenance lies in accurately predicting the future health status and remaining service life of equipment, thereby allowing maintenance to be performed at the optimal time and cost before a failure occurs. Traditional maintenance based on fixed schedules or threshold-based condition monitoring still carries the risk of insufficient or excessive maintenance. Digital twins, by integrating historical data, real-time data, and physical models, make true predictive maintenance possible.

Due to the similarity between the internal insulation of high-voltage bushings and that of transformers, the extensive application of digital twin technology in power equipment operation and maintenance, particularly in transformers and motors, provides many experiences that can be applied to the digital management of bushings. A research team has developed a two-stage life prediction model for generator stator main insulation walls based on digital twins [35]. The study shows that most existing models for predicting the remaining service life of generator main insulation focus on the statistical life distribution of a large number of products. The limitation of these models is that they cannot incorporate individual degradation information of a single product. The model proposed in this study consists of three parts: a common representation model, an individual representation model, and a dynamic evolution model. Its digital twin framework integrates Wiener process models, Kalman filtering algorithms, and support vector machine models, realizing the dynamic evolution of the digital twin system, as shown in Figure 4.

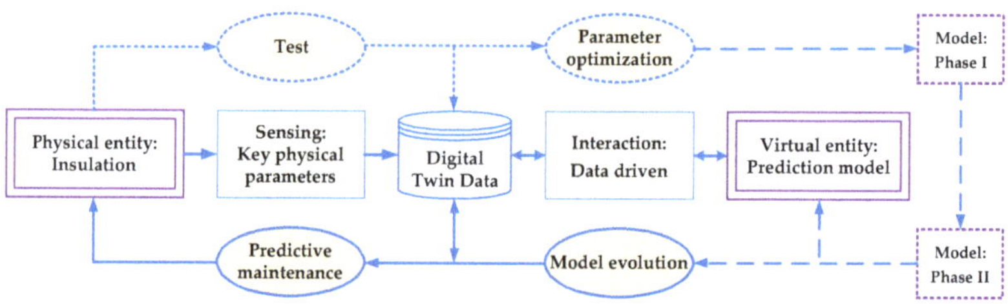

Figure 4. Structure of the digital-twin-driven RUL prediction model. (Indication: This figure is produced by the authors based on the technical logic of Reference [35]).

Predictive maintenance driven by digital twins is a closed-loop optimization process. First, based on the aforementioned intelligent condition assessment model, the system has already obtained the current health index and degradation rate of the equipment. Second, the behavioral prediction model in the twin model (usually a time series prediction model, such as LSTM [56], GRU [57], or Transformer [58]) uses historical time series data combined with future load plans, environmental forecasts, and other information to predict the evolution trajectory for a future period. At the same time, a physics-based degradation model (such as a thermal aging model based on the Arrhenius equation [59], an electrical aging model based on electric field stress [60]) can calculate the accumulated damage under specific operating conditions, providing a physical explanation and constraint for data-driven predictions. When the predicted value is expected to cross a predefined maintenance threshold at some future point, the system will automatically trigger a maintenance decision optimization module. Finally, after the maintenance activities are completed, all data (such as replaced components, repair measures) are fed back into the twin data, realizing model updates and closed-loop iteration. This approach transforms maintenance decisions from "passive response" to "active planning," achieving the best balance between safety and economy.

4. Core Technology System Supporting High-Voltage Bushing Digital Twins

The realization of high-voltage bushing digital twins is not a breakthrough in a single technology, but a complex technology system formed by the deep integration of multidisciplinary and multidomain technologies. This system serves as a bridge connecting physical entities and virtual services, and its advancement directly determines the fidelity, real-time performance, and intelligence level of the digital twin system. The four core technologies involved in high-voltage bushing digital twins are explained as follows: multi-physics field coupling modeling, multi-source and heterogeneous data integration, data-driven model calibration and condition evaluation, and high-performance computing and visualization technologies.

4.1. Multi-Physics Field Coupling Modeling Technology

Multi-physics field coupling modeling is the foundation for constructing high-fidelity virtual models. High-voltage bushings are subjected to complex interactions of electric fields, thermal fields, force fields, and fluid fields (such as oil flow, airflow) during operation. These physical fields interact with and constrain each other, collectively determining the operating state and lifespan of the bushing. For example, dielectric loss and partial discharge generate heat, leading to temperature increases (electro-thermal coupling); temperature changes cause thermal expansion and contraction of materials, generating mechanical stress (thermal-mechanical coupling); and mechanical deformation alters the electric field distribution, exacerbating partial discharge (mechanical-electric coupling) [21,61]. Therefore, simulations of any single physical field cannot truly reflect the internal state of the bushing.

The core of constructing a multi-physics field coupling model lies in solving a system of partial differential equations, with commonly used numerical methods including the finite element method, finite volume method, and boundary element method. In engineering practice, commercial simulation software platforms such as COMSOL Multiphysics and ANSYS are often used [36,45], which provide mature physical field interfaces and coupling solvers. The key to model construction is:

1. The precision of the geometric model, which needs to accurately restore the structure of each layer of insulation, capacitor screens, flanges, etc. of the bushing;
2. The definition of material properties, which requires inputting nonlinear parameters such as dielectric constant, conductivity, thermal conductivity, and elastic modulus that vary with temperature and electric field;
3. The setting of boundary conditions, which needs to accurately apply loads such as voltage, current, and convective heat transfer coefficients.

A high-fidelity coupled model can reproduce the dynamic response of the physical entity under various operating conditions in the digital space, serving as a "virtual testing ground" for subsequent condition evaluation, fault diagnosis, and predictive maintenance.

4.2. Multi-Source Heterogeneous Data Fusion Technology

The vitality of digital twins lies in data. The data generated by high-voltage bushings during operation exhibit typical multi-source heterogeneous characteristics: diverse sources (from various sensors such as DGA, PD, tan δ, temperature, vibration [22,37,62]); diverse types (including numerical data, waveform data, image data, text-based test reports); and diverse temporal and spatial scales (data sampling frequencies range from milliseconds to months). Integrating these isolated, fragmented data into a unified, high-quality information flow is a prerequisite for realizing the value of digital twins.

Multi-source heterogeneous data fusion technology aims to address challenges in data spatiotemporal alignment, noise filtering, feature correlation, and knowledge extraction [23,24]. Its technical system can be divided into three layers:

1. Data layer fusion, primarily focusing on data cleaning, interpolation, normalization, and spatiotemporal alignment to ensure that data from different sources are comparable on a unified spatiotemporal basis;
2. Feature layer fusion, extracting key features from preprocessed data (such as the phase distribution PRPD of PD, the gas ratio features of DGA), and then using feature selection or dimensionality reduction algorithms (such as Principal Component Analysis—PCA) to construct feature vectors that comprehensively reflect the equipment status;
3. Decision layer fusion, using multiple sub-models based on different data sources (such as a fault diagnosis model based on DGA, a defect identification model based on PD) to make preliminary judgments, and then reaching a final comprehensive conclusion through decision fusion algorithms (such as D-S evidence theory, Bayesian networks, weighted voting).

Through multi-level fusion, the digital twin system can overcome the limitations of a single data source, forming a comprehensive understanding of the equipment status that is greater than the sum of its parts.

4.3. Data-Driven Model Calibration and Condition Evaluation Technology

A "static," unchanging virtual model will gradually lose its accuracy as the physical entity ages and wears out, losing its "twin" significance. Therefore, a continuous calibration mechanism must be established to use real-time data from the physical entity to continuously correct the parameters and structure of the virtual model, ensuring that it remains highly consistent with the physical entity. This is data-driven model calibration technology, which is the core feature that distinguishes digital twins from traditional simulations.

Model calibration is typically an optimization problem: defining an objective function (such as the error between simulation output and measured data), and then automatically adjusting the key parameters in the model (such as material conductivity, thermal conductivity, contact resistance) through optimization algorithms (such as Particle Swarm Optimization—PSO [63], Genetic Algorithm—GA [38], Gradient Descent—GD [25]) to minimize the objective function. The calibration process can be offline (using historical data for batch calibration) or online (using real-time data for dynamic updates). The high-fidelity model after calibration, with its simulation output (such as temperature field, electric field distribution), can be considered as "soft measurement" data that cannot be directly measured, greatly enriching the information dimension of condition evaluation. Combined with machine learning algorithms, these "soft measurement" data and direct monitoring data together form the input of the intelligent condition evaluation model, enabling more precise and in-depth evaluation of the health status and evolution trends of the bushing.

4.4. High-Performance Computing and Visualization Technology

The digital twin of high-voltage bushings, especially its multi-physics field coupling model, involves massive computations and large-scale data processing, posing extremely high demands on computational capabilities. At the same time, the complex data generated by the twin system (such as three-dimensional electric field distributions, temperature contour plots, particle trajectories) also require intuitive and efficient presentation methods for operation and maintenance personnel to understand and interact with. High-performance computing and visualization technology are the "engine" and "window" that solve these two problems.

High-performance computing (HPC) aims to significantly enhance computational speed and processing capabilities through parallel computing, distributed computing, and other means [46]. In digital twins, the application of HPC is reflected in:

- Cloud/edge collaborative computing: Complex, computationally intensive multi-physics simulations can be completed on cloud-based high-performance computing clusters, while real-time, lightweight model correction and condition assessment are performed on edge servers close to the equipment, balancing computational depth and response speed.
- Model lightweighting: Using techniques such as reduced-order models and surrogate models, complex finite element models are simplified into mathematical models with minimal computational requirements, enabling them to run in real-time on standard servers or even embedded devices.

Visualization technology is responsible for converting abstract data into intuitive graphs and images [39]. Its applications include:

1. Three-dimensional state visualization: On a three-dimensional model, the internal electric field, temperature, and stress distribution of the bushing are dynamically displayed in the form of contour plots, isocurves, and vector arrows, achieving a "see-through" effect.
2. Augmented Reality (AR)/Virtual Reality (VR) interaction: Operation and maintenance personnel can use AR glasses to overlay virtual temperature field and electric field information onto the real bushing equipment, enabling immersive inspections and fault troubleshooting.
3. Data dashboards: Key indicators such as health index, Remaining Useful Life (RUL) prediction, and alarm information are presented in the form of charts and dashboards, providing decision support for managers.

5. Brief Illustrative Case Study

This paper recently conducted a digital twin application for a 500 kV substation located in Beijing. A digital twin of the substation was constructed, and an effective mapping between the physical entity and its digital counterpart was established through the fusion of sensor networks, operational data, experimental and simulation data, IoT communications, and intelligent algorithmic models. This has enabled the digital and intelligent O&M of the substation based on the digital twin system. The primary implementation pathway and partial visualization outcomes of this project are presented in Figure 5.

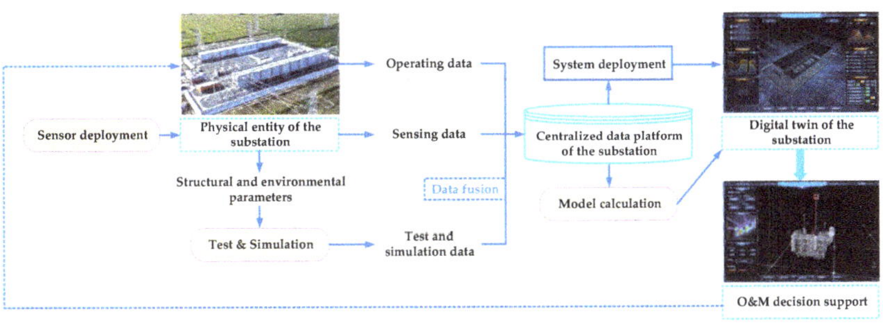

Figure 5. A digital twin application for a 500 kV substation.

6. Challenges and Research Outlook

Today, the world is in a period of rapid information development, with emerging technologies such as advanced sensing, 5G, big data, cloud computing, artificial intelligence, and the Internet of Things (IoT) profoundly impacting the development of various industries. As a representative of cutting-edge information technology, digital twins have gained attention for the digital and intelligent operation and maintenance of power equipment since around 2020 [64–66]. In the research on high-voltage equipment based on digital twins, the main functions that digital twin technology needs to perform and the key technologies involved have generally reached a consensus. The successful implementation of digital twin technology for the FLCM of HV bushings holds immense promise. However, transitioning from conceptual frameworks and academic prototypes to widespread industrial deployment faces significant hurdles.

6.1. Current Research Challenges

Beyond the technical complexities of model fusion and data integration discussed earlier, several practical challenges impede the large-scale adoption of digital twin for HV bushings.

- Data Silos and Heterogeneity: Data forms the foundation of digital twin technology. Data from different sensors (tan δ, PD, temperature, DGA) and systems (e.g., asset management) often exist in proprietary formats and isolated databases. The lack of interoperability makes it difficult to create a unified, coherent data feed for the digital twin. Moreover, as a key component of electrical connections, HV bushings, considering their production costs and issues related to equipment quality and volume, lack high-quality, long-term labeled datasets. This is because power generation and operation departments prioritize safety, leading to a contradiction between data sharing and privacy protection.
- Model credibility and validation: Establishing trust in the digital twin's outputs is paramount. Physics-based models require precise parameters that are often unknown or difficult to measure in-situ. Data-driven models, particularly deep learning, can act as "black boxes," making it hard to interpret their predictions. Rigorous validation against long-term operational data and failure cases is often lacking. In addition, for the simulation model, there are still some technical details to overcome in multi-physics and coupled modeling, as well as real-time simulation analysis. Unified modeling across multiple scales and mechanisms remains a challenge, which affects the interpretability and credibility of the models.
- Computational and infrastructure costs: Running high-fidelity, multi-physics simulations in near real-time requires substantial computational resources. This poses a challenge for edge deployment in substations with limited IT infrastructure and necessitates a careful balance between model accuracy and computational efficiency.
- Lack of standardized frameworks: The absence of industry-wide standards for digital twin architecture, data semantics, and communication protocols leads to vendor-specific, non-interoperable solutions, increasing integration costs and locking utilities into single-vendor ecosystems. An effective mapping relationship needs to be established between the physical entity and the digital twin. Currently, there is a lack of unified data interfaces, model formats, and functional standards, making it difficult for systems to interconnect, especially between digital twins of different equipment. Effective communication between digital twins still needs to be achieved based on the electrical interconnection of physical entities.
- The return on investment (ROI) is unclear: In terms of operational costs and engineering applications, digital twin technology requires the deployment of systematic high-precision sensors. Considering the economic efficiency of equipment lifecycle management, the current development, deployment, and maintenance costs of twin systems are relatively high, and the return on investment (ROI) is not yet clear.

6.2. Future Research Directions

In response to the aforementioned challenges, future research on the full lifecycle management of high-voltage bushings can be conducted from the following aspects.

6.2.1. Standardization and Interoperability

The future of bushing DTs hinges on the adoption of open standards to ensure seamless data exchange and model portability. Standardized data models and interfaces are of great significance for digital twin technology to better serve the O&M of high-voltage equipment.

- IEC 60137 & IEC 61439 [67,68]: These standards define the specifications and tests for HV bushings. Future extensions or companion standards could formally define a standardized "digital twin information model" for bushings, specifying a minimum set of parameters, data points (e.g., tan δ, PD, temperature profiles), and performance metrics to be exchanged.
- IEC 61850 & CIM [69,70]: For communication within the substation and enterprise levels, strict adherence to IEC 61850 (for real-time data exchange) and the Common Information Model (CIM) (for asset management) is crucial. Research should focus on defining new Logical Nodes within the IEC 61,850 framework to standardize how bushing digital twin data is represented and accessed.
- OPC UA [71]: As a platform-independent, service-oriented communication standard, OPC UA is ideal for bridging the gap between the substation floor and enterprise IT systems. Its companion specifications can model the asset hierarchy and production processes, providing a robust backbone for DT data contextualization.

To fully leverage the functions of the digital twin system, the development of relevant standards should be promoted, and an open digital twin platform and ecosystem should be constructed to lower the threshold for the application of digital twin lifecycle management systems.

6.2.2. System Deployment in Response to Actual Situations

The design of a reliable distributed system must comprehensively address the practical constraints of its operational environment from the outset. Within the context of edge computing, these systems constitute high-value cyber-physical assets, making the security of the end-to-end data pipeline—from sensors to the cloud—a critical imperative. To this end, future research must prioritize several key areas. These include the implementation of a defense in depth strategy through network segmentation, firewalls, and intrusion detection systems; the utilization of technologies like digital signatures and blockchain to ensure data integrity and authenticity while providing an immutable ledger for critical events; and the adoption of a "Zero Trust" architecture, which operates on the principle of never trusting and always verifying, regardless of a device's or user's network location.

Regarding the advancement of sensor calibration, traceability, and condition monitoring within digital twin systems, the next generation of "smart sensors" must possess inherent self-diagnostic and condition monitoring functionalities. Such sensors are required to autonomously report their operational status and potential deviations to the data acquisition system. This enables the data acquisition system to either mitigate the impact of unreliable data in real-time or generate corresponding alerts. Moreover, to reconcile the competing requirements of real-time responsiveness and high-fidelity simulation, a hybrid edge-cloud computing architecture emerges as the most viable solution, offering significant optimization of bandwidth utilization and computational expenditure.

6.2.3. Economic Viability and ROI Modeling

To incentivize utility companies to invest in digital technologies, a clear and compelling business case is paramount. Future research must move beyond qualitative descriptions of technical benefits and focus on developing quantitative ROI models. These models should be characterized by low cost, low risk, and high efficiency. This can be achieved by extending asset lifespans to defer capital expenditures, quantifying the probability and associated costs of catastrophic failures, and ultimately improving maintenance schedules to reduce required manpower and downtime. Concurrently, future research should concentrate on value-based data acquisition strategies to determine the "minimum sufficient dataset" required for achieving a specific predictive accuracy, thereby enabling a phased, low-cost deployment of digital technologies.

6.2.4. System Deployment and Technology Integration

A key future direction involves the evolution from digital twins of individual devices to group digital twins. By constructing a group digital twin for high-voltage bushings within a substation or even a regional power grid, the interdependencies and synergistic O&M of the equipment can be systematically analyzed.

Beyond this architectural evolution, deeper integration with AI is imperative. Future research should explore the application of large language models (LLMs) and reinforcement learning (RL) to enable intelligent interaction, autonomous decision-making, and optimal control within the bushing digital twin management system.

From a systems engineering perspective, the deployment of these twins should adopt a collaborative "cloud-edge-end" architecture. This necessitates research into deploying lightweight models at the edge to ensure rapid response and real-time control, while the cloud platform handles complex computations and global optimization tasks.

Furthermore, the dynamism and openness of the new power system and its associated electricity market must be considered. Consequently, digital twin systems should be leveraged to support the broader power energy system across all sectors—generation, transmission, distribution, and sales. In this context, exploration can be directed towards equipment asset value assessment, green O&M certification, and carbon footprint tracking based on twin data, ultimately contributing to the global carbon emissions equilibrium.

7. Conclusions

High-voltage bushings, as a critical hub ensuring the safe and stable operation of power transmission and transformation systems, require an intelligent transformation of their management model to build a new power system. This paper reviews the research on the full lifecycle management of high-voltage bushings based on digital twins, aiming to provide theoretical references and technical guidance for overcoming the bottlenecks of traditional management models and enhancing equipment operation and maintenance levels.

This paper first constructs a five-dimensional digital twin management framework that integrates design, manufacturing, operation, and retirement stages, providing a top-level design for the digital and transparent management of high-voltage bushings. On this basis, it deeply analyzes the application paradigms of digital twins in typical scenarios such as external insulation design optimization, intelligent condition assessment, insulation defect identification, fault diagnosis, and predictive maintenance, revealing their revolutionary transformation from passive response to active prevention and from empirical decision-making to data-driven. Furthermore, this paper reviews the core technology systems supporting these applications, such as multi-physics field coupling modeling, multi-source heterogeneous data fusion, and data-driven model correction, clarifying the technical paths for achieving high-fidelity virtual-physical mapping and intelligent decision-making.

However, the in-depth application of digital twin technology in the field of high-voltage bushings still faces multiple challenges, including data quality, model credibility, lack of standard systems, and high engineering costs. Looking forward, we believe that building a group of digital twins for substation equipment, deepening the integration with artificial intelligence, exploring the cloud-edge-end collaborative deployment architecture, and accelerating the construction of industry standards and ecosystems will be key directions for promoting this technology from theory to large-scale application.

In summary, digital twin technology is not only an effective means to improve the intelligent operation and maintenance level of high-voltage bushings but also a core engine to drive the digitalization of power equipment asset management across all elements and the entire lifecycle, having profound strategic significance for achieving the goals of "carbon peak and carbon neutrality".

Author Contributions: Conceptualization, W.C. and C.Z.; methodology, T.W., J.Z. and Z.W.; validation, W.C., T.W. and J.Z.; formal analysis, W.C.; investigation, Z.W.; resources, W.C. and C.Z.; data curation, T.W. and J.Z.; writing—original draft preparation, Z.W.; writing—review and editing, C.Z.; project administration, W.C., T.W. and J.Z.; funding acquisition, C.Z. All authors have read and agreed to the published version of the manuscript.

Funding: This research was funded by Science and Technology Project of State Grid Hebei Electric Power Co., Ltd., grant number SGHEXA00YJJS2400654.

Data Availability Statement: No new data were created or analyzed in this study.

Acknowledgments: The authors gratefully acknowledge the contributions of Zinan Shi and Aichun Sun of Syi Tsing Energy Tech. and Yang Zhan of North China Electric Power University for their help on the previous research of this paper.

Conflicts of Interest: Authors Weiwei Chi, Tao Wang and Jichao Zhang were employed by the company State Grid Hebei Electric Power Co., Ltd. The remaining authors declare that the research was conducted in the absence of any commercial or financial relationships that could be construed as a potential conflict of interest.

Abbreviations

The following abbreviations are used in this manuscript:

HV	High voltage
FLCM	Full life cycle management
O&M	Operation and maintenance
HVDC	High voltage direct current
CBM	Condition-based maintenance
RIP	Resin-impregnated paper
RIS	Resin-impregnated synthetic
PD	Partial discharge
RUL	Remaining useful life
RNNs	Recurrent neural networks
LSTM	Long short-term emory network
IoT	Internet of things
GA	Genetic algorithm
PSO	Particle swarm optimization
DGA	Dissolved gas analysis
CNN	Convolutional neural network

GRU	Gated recurrent unit
PCA	Principal component analysis
GD	Gradient descent
HPC	High performance computing
AR	Augmented Reality
VR	Virtual reality
ROI	Return on investment
IT	Information technology
CIM	Common information model
LLMs	Large language models
RL	Reinforcement learning

References

1. Wei, Y.M.; Chen, K.Y.; Kang, J.N.; Chen, W.M.; Zhang, X.Y.; Wang, X.Y. Policy and Management of Carbon Peaking and Carbon Neutrality: A Literature Review. *Engineering* **2022**, *14*, 52–63. [CrossRef]
2. Du, X.W. Thoughts on strategies and paths to achieve carbon peaking and carbon neutrality in China. *Front. Energy* **2023**, *17*, 324–331. [CrossRef]
3. Sun, J.Y.; Su, C.G.; Song, J.C.; Yao, C.C.; Ren, Z.M.; Sui, Q. Capacity planning for large-scale wind-photovoltaic-pumped hydro storage energy bases based on ultra-high voltage direct current power transmission. *Energy* **2025**, *320*, 135224. [CrossRef]
4. Glende, E.; Wolter, M. Extra-long distance ultra-high voltage direct current. In Proceedings of the 2022 IEEE Sustainable Power and Energy Conference, Perth, Australia, 4–7 December 2022; pp. 1–5. [CrossRef]
5. Christin, A.J.; Salam, M.A.; Rahman, Q.M.; Islam, M.A.; Wen, F.S.; Ang, S.P.; Hasan, S.; Voon, W. Investigation of failure of high voltage bushing at power transformer. *J. Electrost.* **2018**, *96*, 49–56. [CrossRef]
6. Li, X.; Tang, H.; Yang, F.; Cheng, H.; Zhang, S. Internal Discharge Characteristics of HVDC Wall Bushing Model with Resin Impregnated Paper Capacitor Core under Long-term DC Voltage of Different Polarity. In Proceedings of the 2022 IEEE International Conference on High Voltage Engineering and Applications, Chongqing, China, 25–29 September 2022; pp. 1–4. [CrossRef]
7. Fink, S.; Müller, R.; Zwecker, V. Cryogenic high voltage testing of a 25 kV RIS bushing. In Proceedings of the 2022 IEEE International Conference on High Voltage Engineering and Applications, Chongqing, China, 25–29 September 2022; pp. 1–4. [CrossRef]
8. Go, M.Y.L.; Chan, K.K.M.; Wong, S.C.H.; Chan, J.T.J. Life management of high voltage bushings. In Proceedings of the 10th International Conference on Advances in Power System Control, Operation & Management, Hong Kong, China, 8–12 November 2015; pp. 1–5. [CrossRef]
9. Grieves, M. Digital Twin: Manufacturing Excellence Through Virtual Factory Replication. 2015. Available online: https://www.researchgate.net/publication/275211047_Digital_Twin_Manufacturing_Excellence_through_Virtual_Factory_Replication (accessed on 15 March 2015).
10. Tao, F.; Zhang, M.; Nee, A.Y.C. *Digital Twin Driven Smart Manufacturing*; Elsevier: Amsterdam, The Netherlands, 2019; pp. 1–257.
11. Tao, F.; Zhang, H.; Liu, A.; Nee, A.Y.C. Digital Twin in Industry: State-of-the-Art. *IEEE Trans. Ind. Inform.* **2019**, *15*, 2405–2415. [CrossRef]
12. Wang, T.X.; Liu, Z.Y.; Wang, Z.Z.; Liu, J.; Matharage, S.Y.; Wang, Z.D. Transformer Digital Twin Development Framework and Functionality Demonstration. In Proceedings of the 2024 IEEE International Conference on High Voltage Engineering and Applications, Berlin, Germany, 18–22 August 2024. [CrossRef]
13. Tao, F.; Zhang, M. Digital Twin Shop-Floor: A New Shop-Floor Paradigm Towards Smart Manufacturing. *IEEE Access* **2017**, *5*, 20418–20427. [CrossRef]
14. Zhang, S.L.; Zhang, S.Y.; Su, X.L. Multi-Physics Field Simulation of 1100kV GIS SF6 Gas Insulated Composite Bushing. In Proceedings of the 2019 IEEE Conference on Electrical Insulation and Dielectric Phenomena, Richland, WA, USA, 20–23 October 2019; pp. 247–250. [CrossRef]
15. He, Y.; Zhang, S.; Wang, P.; Dai, L. Digital Twin Model of Converter Transformer Output Device Based on Real-time Inversion Calculation of Transient Temperature and E-field of HVDC Bushing. In Proceedings of the 2024 International Conference on Artificial Intelligence, Deep Learning and Neural Networks, Guangzhou, China, 20–22 September 2024; pp. 57–64. [CrossRef]
16. Wu, Z.; La, Y.; Wang, Z.; Zhang, W.; Cheng, J.; Huang, K. A Real-Time Simulation Model for the Digital Twin of the DC Transformer Bushing. In Proceedings of the 2024 Boao New Power System International Forum—Power System and New Energy Technology Innovation Forum, Qionghai, China, 8–10 December 2024; pp. 483–487. [CrossRef]
17. Zhang, L.; Sun, L.; Wu, J.F.; Han, Y.H.; Wang, S.; Yang, C.K.; Shen, W.; Guo, C. Development of multi-parameter online monitoring equipment for EHV transformer bushing. *IET Sci. Meas. Technol.* **2020**, *14*, 89–103. [CrossRef]

18. Subocz, J.; Mrozik, A.; Bohatyrewicz, P.; Zenker, M. Condition Assessment of HV Bushings with Solid Insulation based on the SVM and the FDS Methods. *Energies* **2020**, *13*, 853. [CrossRef]
19. Zhang, S.L.; Deng, B.J.; He, Y.S.; Dai, L.J. Research on Structural Specification and Virtual Construction Method of Physical -Field for Digital Twin Model of C5F10O Insulated Bushing. In Proceedings of the 2024 14th International Conference on Power and Energy Systems, Chengdu, China, 13–16 December 2024; pp. 160–166. [CrossRef]
20. Zhang, S.L.; Wang, P.B.; Li, X.C.; Xiao, Q. Dielectric Spectrum and Insulation State Analysis of Ultra-High Voltage Converter Transformer Bushing Based on Small Sample Digital Twin Model. In Proceedings of the 2025 8th International Conference on Electronics Technology, Chengdu, China, 17–20 May 2025; pp. 320–326. [CrossRef]
21. Zhang, S.L.; Xiao, Q.; Wu, J.G.; Ye, F. Electrical and Insulation Characteristics of Digital Twin Model of Valve Side Bushing for Ultra-high Voltage Converter Transformer Based on Field circuit Coupling and Transient Lightning Impulse Simulation. In Proceedings of the 2025 International Conference on Smart Grid and Sustainable Energy, Zhangye, China, 25–27 July 2025; pp. 35–41. [CrossRef]
22. Sun, Y.; Shi, Y.; Hu, Q.; Xie, C.; Su, T. DTformer: An Efficient Digital Twin Model for Loss Measurement in UHVDC Transmission Systems. *IEEE Trans. Power Syst.* **2024**, *39*, 3548–3559. [CrossRef]
23. Zhang, J.; Liu, X.; Wu, J.; Wu, X. Research on Key Technology of Digital Twin of Transformer Equipment for Technology Compliance Assessment. In Proceedings of the 2025 10th Asia Conference on Power and Electrical Engineering, Beijing, China, 15–19 April 2025; pp. 6–10. [CrossRef]
24. Moutis, P.; Alizadeh-Mousavi, O. Digital Twin of Distribution Power Transformer for Real-Time Monitoring of Medium Voltage from Low Voltage Measurements. *IEEE Trans. Power Deliv.* **2021**, *36*, 1952–1963. [CrossRef]
25. He, Y.S.; Li, Y.F.; Li, X.C.; Yuan, Y.N.; Yang, F.; Lu, Z.X. A Reduced-Order Algorithm for a Digital Twin Model of Ultra-High-Voltage Valve-Side Bushing Considering Spatio-Temporal Non-Uniformity. *Energies* **2025**, *18*, 1481. [CrossRef]
26. Ding, F.; Ma, Y.; Li, X.; Huang, J.; Kang, J. Research on digital twin diagnosis model for the thermal-electric field of high-voltage switchgears. *Sci. Rep.* **2025**, *15*, 34303. [CrossRef] [PubMed]
27. Liu, Y.; Nie, K.; Li, C.; Li, Z.; Gao, W.; Meng, F.; Li, Q. Electric Field Simulation Analysis of Typical Defects of Metal Particles in Main Insulation of Converter Transformer Based on Multi-physics Coupling. In Proceedings of the 5th International Symposium on Plasma and Energy Conversion, Singapore, 27–29 October 2023; Springer Proceedings in Physics; Volume 398, pp. 449–459. [CrossRef]
28. Mohammadnabi, S.; Rahmani, K. Influence of humidity and contamination on the leakage current of 230-kV composite insulator. *Electr. Power Syst. Res.* **2021**, *194*, 107083. [CrossRef]
29. Li, X.L.; Wan, M.D.; Zhang, G.K.; Lin, X. Surface Charge Characteristics of DC-GIL Insulator Under Multiphysics Fields: 3-D Modeling. *IEEE Trans. Dielectr. Electr. Insul.* **2022**, *29*, 1608–1616. [CrossRef]
30. Zhou, Z.; Li, H.; Wen, S.; Zhang, C. Prediction Model for the DC Flashover Voltage of a Composite Insulator Based on a BP Neural Network. *Energies* **2023**, *16*, 984. [CrossRef]
31. Zhang, L.; Sun, L.; Wang, W.S.; Han, Y.H.; Pu, L.; Wu, J.F.; Wu, H. Intelligent monitoring of EHV transformer bushing based on multi-parameter composite sensing technology. *IET Sci. Meas. Technol.* **2023**, *17*, 373–384. [CrossRef]
32. Mitiche, I.; McGrail, T.; Boreham, P.; Nesbitt, A.; Morison, G. Data-Driven Anomaly Detection in High-Voltage Transformer Bushings with LSTM Auto-Encoder. *Sensors* **2021**, *21*, 7426. [CrossRef]
33. Zhang, S.; Xiao, Q.; Wu, J.; Ye, F. Implementation and Dynamic Online Application of Digital Twin Industrial Simulation Platform for Valve Side Bushing of Converter Transformer Based on Multi Physics Coupling Technology. In Proceedings of the 2025 International Conference on Smart Grid and Sustainable Energy, Zhangye, China, 25–27 July 2025; pp. 47–53. [CrossRef]
34. Zhang, S.; Xiao, Q.; Wu, J.; Yang, F. Research on Real Time Configuration and Evaluation Service of Ultra high Voltage Valve Side Bushing Driven by Multi Source Sensor Data and Material Parameters. In Proceedings of the 2025 International Conference on Smart Grid and Sustainable Energy, Zhangye, China, 25–27 July 2025; pp. 7–13. [CrossRef]
35. Zhang, Q.; Wu, J.; Wang, J.; Huang, X.; Fang, Y.; Niu, F.; Zhang, J. A Two-Phase Lifetime Prediction Model of Generator Stator Main Wall Insulation Driven by Digital Twin. *IEEE Trans. Instrum. Meas.* **2024**, *73*, 3531512. [CrossRef]
36. Bai, F.; Tang, Z.; Yin, R.J.; Quan, H.B.; Chen, L.; Dai, D.; Tao, W.Q. A novel '3D+digital twin+3D' upscaling strategy for predicting the detailed multi-physics distributions in a commercial-size proton exchange membrane fuel cell stack. *Appl. Energy* **2024**, *374*, 124012. [CrossRef]
37. Sun, L.M.; Sun, H.B.; Zhang, W.; Li, X.Y. Hybrid monitoring methodology: A model-data integrated digital twin framework for structural health monitoring and full-field virtual sensing. *Adv. Eng. Inform.* **2024**, *60*, 102386. [CrossRef]
38. Huang, X.J.; Lang, X.; Guo, T.; Yu, L. Improving pyraformer algorithm for forecasting of 500 KV transformer bushing data with three enhanced modules. *Electr. Power Syst. Res.* **2025**, *241*, 111360. [CrossRef]
39. Li, Z.; Liu, K.; Lin, M.; Xin, D.L.; Tang, H.; Wu, G.N. A zero-sample state evaluation model for valve-side bushing of UHV converter transformer oriented to digital twin under attribute analysis. *IET Gener. Transm. Distrib.* **2023**, *17*, 1123–1134. [CrossRef]

40. Zhang, L.; Liu, J.H.; Zhuang, C.B. Digital Twin Modeling Enabled Machine Tool Intelligence: A Review. *Chin. J. Mech. Eng.* **2024**, *37*, 47. [CrossRef]
41. Tao, F.; Liu, A.; Hu, T.L.; Nee, A.Y.C. *Digital Twin Driven Smart Design*; Elsevier: Amsterdam, The Netherlands, 2020. [CrossRef]
42. Tao, F.; Zhang, M.; Cheng, J. Digital twin workshop: A new paradigm for future workshop. *Comput. Integr. Manuf. Syst.* **2017**, *23*, 1–9. [CrossRef]
43. Tao, F.; Cheng, J.; Qi, Q.; Zhang, M.; Zhang, H.; Sui, F. Digital twin-driven product design, manufacturing and service with big data. *Int. J. Adv. Manuf. Technol.* **2018**, *94*, 3563–3576. [CrossRef]
44. Barricelli, B.R.; Casiraghi, E.; Fogli, D. A Survey on Digital Twin: Definitions, Characteristics, Applications, and Design Implications. *IEEE Access* **2019**, *7*, 167653–167671. [CrossRef]
45. Lehner, D.; Pfeiffer, J.; Tinsel, E.F.; Strljic, M.M.; Sint, S.; Vierhauser, M. Digital Twin Platforms: Requirements, Capabilities, and Future Prospects. *IEEE Softw.* **2022**, *39*, 53–61. [CrossRef]
46. Maiterth, M.; Brewer, W.; Vet, D.D.; Greenwood, S.; Kumar, V.; Hines, J. Visualizing an Exascale Data Center Digital Twin: Considerations, Challenges and Opportunities. In Proceedings of the 2024 IEEE Visualization and Visual Analytics, St. Pete Beach, FL, USA, 13–18 October 2024; pp. 21–25. [CrossRef]
47. Shunmugalakshmi, G.; Maheswari, R.V.; Vigneshwaran, B. A Comprehensive Study and Prediction of Flashover Voltage of 11kV Bushing with Various Pollutions using Support Vector Machine. In Proceedings of the IEEE International Conference on Innovations in Power and Advanced Computing Technologies, Vellore, India, 22–23 March 2019. [CrossRef]
48. Qi, B.; Yang, Z.; Yang, X.; Gao, C.; Lu, L.; Zhang, Y. Effect of Different DC Prestressed Time on Flashover Characteristics of Epoxy Resin Under Polarity Reversal Voltage. *IEEE Trans. Dielectr. Electr. Insul.* **2022**, *29*, 1582–1589. [CrossRef]
49. Zhang, Z.; Zhang, Y.; Xiao, W.; Sun, Y.; Jiang, X. Research on Insulation Configuration of 220 kV Bushing in High-Altitude Polluted Areas. *Energies* **2024**, *17*, 1645. [CrossRef]
50. Li, W.; Wang, Y.; Ding, Y.; Yin, Y. Optimization Design of Packaging Insulation for Half-Bridge SiC MOSFET Power Module Based on Multi-Physics Simulation. *Energies* **2022**, *15*, 4884. [CrossRef]
51. Wu, Z.; Xin, L.; Cheng, J.; Wu, B.; Qiang, Z.; Wang, Q.; Zhao, L.; Du, Z. Research on the Data-Driven Fast Calculation Method of the Temperature Field Distribution of Valve-Side RIP Bushing Used in UHV DC Converter Transformer. *High Volt.* **2025**, *10*, 1210–1220. [CrossRef]
52. Bueno-Barrachina, J.M.; Ye-Lin, Y.; Nieto-del-Amor, F.; Fuster-Roig, V. Inception 1D-convolutional neural network for accurate prediction of electrical insulator leakage current from environmental data during its normal operation using long-term recording. *Eng. Appl. Artif. Intell.* **2023**, *119*, 105799. [CrossRef]
53. Han, T.; Hao, C.; Gao, Y.; Du, B.; Cavallini, A. Partial Discharge Detection with Bushing Sensors for Switchgear: Separation, Location, and Recognition. *IEEE Electr. Insul. Mag.* **2025**, *41*, 5–14. [CrossRef]
54. Tian, T.; Zhou, X.; Dai, L.C.; Bai, J.; Yang, W.D.; Luo, Y.; Pan, L.L.; Yu, J.Y.; Li, J.B.; Chang, X.H.; et al. Research on fatigue fracture failure of high-voltage bushing conductive rods under complex conditions. *AIP Adv.* **2025**, *15*, 095127. [CrossRef]
55. Jiang, J.; Bie, Y.; Li, J.; Yang, X.; Ma, G.; Lu, C.; Zhang, C. Fault diagnosis of the bushing infrared images based on mask R-CNN and improved PCNN joint algorithm. *High Volt.* **2021**, *6*, 116–124. [CrossRef]
56. Wen, X.Y.; Li, W.B. Time Series Prediction Based on LSTM-Attention-LSTM Model. *IEEE Access* **2023**, *11*, 48322–48331. [CrossRef]
57. Qiu, B.; Zhang, J.; Yang, Y.; Qin, G.; Zhou, Z.; Ying, C. Research on Oil Well Production Prediction Based on GRU-KAN Model Optimized by PSO. *Energies* **2024**, *17*, 5502. [CrossRef]
58. Feng, J.; Feng, Z.; Jiang, G.; Zhang, G.; Jin, W.; Zhu, H. A Prediction Method for the Average Winding Temperature of a Transformer Based on the Fully Connected Neural Network. *Appl. Sci.* **2024**, *14*, 6841. [CrossRef]
59. Yin, S.; Bai, Y.; Kong, F.; Wang, Z.; Fang, C. An Interval Prediction of Chloroprene Rubber Crack Propagation Characteristics Based on Thermal Accelerated Aging. *Polymers* **2023**, *15*, 2445. [CrossRef]
60. Halloum, M.R.; Reddy, B.S.; Reddy, G.N. Failure analysis of field-aged polymeric outdoor insulators and performance enhancement for electric stress using nonlinear field grading composites. *Electr. Eng.* **2024**, *106*, 1147–1161. [CrossRef]
61. Chu, Z.L.; Wang, Q.Y.; Tian, H.D.; Liu, S.Y.; Gao, Z.J.; Tian, S.B.; Liu, P.; Peng, Z.R. Fast computation of electro-thermal coupling field of ± 400 kV converter transformer valve side bushing based on proper orthogonal decomposition method. *Int. J. Electr. Power Energy Syst.* **2025**, *167*, 110617. [CrossRef]
62. Zhu, J.X.; Yang, Y.X.; Xi, M.X.; Ji, S.L.; Jia, L.Y.; Hu, T. The next-generation digital twin: From advanced sensing towards artificial intelligence-assisted physical-virtual system. *J. Ind. Inf. Integr.* **2025**, *48*, 100942. [CrossRef]
63. Li, P.; Peng, X.; Yin, K.; Xue, Y.; Wang, R.; Ma, Z. 3D Localization Method of Partial Discharge in Air-Insulated Substation Based on Improved Particle Swarm Optimization Algorithm. *Symmetry* **2022**, *14*, 1241. [CrossRef]
64. Beheshti Asl, M.; Fofana, I.; Meghnefi, F. Review of Various Sensor Technologies in Monitoring the Condition of Power Transformers. *Energies* **2024**, *17*, 3533. [CrossRef]
65. Liu, L.; Guo, Y.; Yin, W.; Lei, G.; Zhu, J. Design and Optimization Technologies of Permanent Magnet Machines and Drive Systems Based on Digital Twin Model. *Energies* **2022**, *15*, 6186. [CrossRef]

66. Ibrahim, M.; Rassõlkin, A.; Vaimann, T.; Kallaste, A.; Zakis, J.; Hyunh, V.K.; Pomarnacki, R. Digital Twin as a Virtual Sensor for Wind Turbine Applications. *Energies* **2023**, *16*, 6246. [CrossRef]
67. *IEC 60137*; Bushings for Alternating Voltages Above 1000 V. International Electrotechnical Commission: Geneva, Switzerland, 2018.
68. *IEC 61439-1*; Low-Voltage Switchgear and Control Gear Assemblies—Part 1: General Rules. International Electrotechnical Commission: Geneva, Switzerland, 2021.
69. *IEC 61850*; Communication Networks and Systems for Power Utility Automation. International Electrotechnical Commission: Geneva, Switzerland, 2013.
70. *IEC 61970*; Energy Management System Application Program Interface (EMS-API)—Part 301: Common Information Model (CIM) base. International Electrotechnical Commission: Geneva, Switzerland, 2020.
71. OPC Foundation. *OPC Unified Architecture Specification, Part 1: Overview and Concepts, Release 1.05*; OPC Foundation: Scottsdale, AZ, USA, 2023.

MDPI AG
Grosspeteranlage 5
4052 Basel
Switzerland
Tel.: +41 61 683 77 34

Energies Editorial Office
E-mail: energies@mdpi.com
www.mdpi.com/journal/energies

www.ingramcontent.com/pod-product-compliance
Lightning Source LLC
Chambersburg PA
CBHW061555210926
51850CB00009B/2112
* 9 7 8 3 7 2 5 8 6 9 7 2 5 *